Secrets of the Cosmos

SINYORITA B.

SECRETS OF THE COSMOS

FIFTH EDITION

2024

Secrets of The Cosmos

Sinyorita B.

ISBN 9798880083640

Preface

Earth, a tiny and dazzling speck in the vast cosmic canvas, is not just our home but a cradle of human curiosity, scientific exploration, and boundless wonder. Inhabiting this small planet, we are not mere witnesses to the grandeur of the universe; we are intrepid explorers who tirelessly venture beyond our terrestrial boundaries. Our quest leads us from the enigmatic birth and expansion of the universe to the hunt for habitable worlds and the complexities of interstellar communication. It reminds us of the ethical considerations surrounding potential encounters with extraterrestrial life and the responsible use of cosmic resources. This book encapsulates the captivating narrative of our journey, revolving around the fundamental question: "Secrets of the Cosmos"

The odyssey begins with a profound exploration of the birth and expansion of the universe, an event that unfolds from the cataclysmic Big Bang to the formation of galaxies, stars, and planets. Yet, our journey extends far beyond our planetary confines. It takes us to the tantalizing Goldilocks zone, where the conditions for life may exist beyond Earth. We peer into the vast expanse of exoplanets, each with its unique story waiting to be uncovered.

As cosmic pioneers, we confront the intricacies of interstellar communication. We send signals into the cosmic void, hoping for a response that bridges the unfathomable distances separating us from potential extraterrestrial civilizations. These exchanges, transcending our imagination, symbolize the pinnacle of our pursuit.

Our expedition is not an individual undertaking but a collective odyssey uniting us as seekers of truth, understanding, and communion with the cosmos. As we traverse the limitless expanse of space, we recognize that our exploration is an eternal journey—a testament to the boundless human spirit and an enduring tribute to our intrinsic desire to comprehend the universe enveloping us.

Thank you for embarking on this journey with us. Now, let us together dive into the depths of the cosmos, explore its mysteries, and seek answers to the questions that propel us forward on this enchanting adventure.

Contents

Introduction

1.1. The Vastness of the Universe and the Unknown

1.1.1. The Birth and Expansion of the Universe

The birth and expansion of the universe constitute one of the most fundamental topics in modern cosmology. The universe came into existence through an event known as the Big Bang. The foundation of this event lies in the concept of a singularity—a moment where everything was concentrated into a single point—followed by an explosion, essentially the "zero moment." Immediately after the explosion, the universe began its expansion, a process that continues to this day.

This expansion is an observed phenomenon throughout the universe. Recent observations indicate that the rate of expansion of the universe is accelerating, a discovery that remains a profound mystery for cosmologists. This rapid expansion appears to be driven by a mysterious form of energy known as dark energy.

This expansion reveals that the universe was once much hotter and denser just a few billion years ago. During its early moments, the universe's temperature and pressure were incredibly high, providing an ideal environment for the formation of fundamental particles. Therefore, the universe's inception is regarded as a captivating topic that transcends both time and space.

Moreover, the birth and expansion of this universe laid the foundation for subsequent evolutionary processes, including the formation of black holes, stars, galaxies, and planets. Understanding the age and structure of the universe remains a significant area of research in cosmology and astrophysics, considered one of the most captivating questions in modern science.

Comprehending this process is an ongoing endeavor for astronomers and physicists, with continuous advancements in telescopes and observation equipment enabling deeper exploration. While the universe's origin and expansion still hold significant mysteries, the investigations in this field continue to be a source of great excitement within the scientific community.

1.1.2. Grasping Cosmic Dimensions

Grasping the vast dimensions of the cosmos is a humbling endeavor that challenges the very limits of human understanding. The universe, with its billions of galaxies, each containing billions of stars and potentially even more planets, presents an unimaginable scale.

The concept of cosmic dimensions extends beyond just the sheer number of celestial objects. It encompasses the enormous distances that separate them. Light, which travels at an astonishing speed of approximately 299,792 kilometers per second (or about 186,282 miles per second) – a value denoted as "c" in physics – still takes years, centuries, or even billions of years to traverse the distances between cosmic entities.

The term "light-year" is often used to describe these immense distances. A light-year is the distance that light travels in one year, approximately 9.46 trillion kilometers (about 5.88 trillion miles). When we speak of distant galaxies or the nearest stars beyond our solar system, we often use light-years as a unit of measurement.

Grasping the cosmic dimensions requires not only a deep understanding of mathematics and physics but also an ability to contemplate scales that are far beyond our everyday experiences. It invites us to ponder the immense voids of interstellar space, the colossal size of galaxies, and the intricate dance of celestial bodies governed by the laws of gravity.

As we delve deeper into the cosmos, we come to appreciate the unfathomable vastness of the universe, where the light from distant stars may have traveled for billions of years to reach our eyes. The study of these cosmic dimensions fuels our curiosity and drives us to explore the mysteries that lie beyond our terrestrial boundaries.

1.1.3. The Allure of the Unknown

The allure of the unknown has always been a driving force behind human curiosity and exploration. When it comes to the cosmos, this fascination reaches new heights. The vastness of the universe, its mysteries, and its uncharted territories have an irresistible pull on our imagination.

One of the most compelling aspects of the unknown in the universe is the prospect of discovering extraterrestrial life. Are we alone in the cosmos, or are there other intelligent beings out there? This question has captivated scientists, philosophers, and dreamers alike for centuries. The pursuit of this answer has led to groundbreaking discoveries in astronomy and astrobiology.

The allure of the unknown also extends to the enigmatic phenomena that occur in the universe. From the perplexing behavior of dark matter to the elusive nature of black holes, the cosmos is riddled with puzzles waiting to be solved. Each mystery presents an opportunity for scientists to push the boundaries of human knowledge.

Exploring the unknown in space requires cutting-edge technology and innovative thinking. Telescopes, space probes, and observatories have become our tools for unraveling cosmic mysteries. They allow us to peer into the depths of the universe, providing glimpses of distant galaxies, cosmic explosions, and the birth and death of stars.

Moreover, the allure of the unknown is not limited to the scientific community. It captures the imagination of the public, inspiring countless works of science fiction, art, and literature. It fuels our desire to venture beyond our home planet, fostering dreams of human colonization of other worlds.

As we continue to probe the depths of space, we are driven by the allure of the unknown, pushing the boundaries of human knowledge and expanding our understanding of the cosmos. It is this relentless curiosity and the quest for the undiscovered that propel us into the vast and mysterious universe that surrounds us.

1.2. The Loneliness Conundrum in the Universe

1.2.1. Fermi Paradox: Where Are They?

The Fermi Paradox is a thought-provoking question that has puzzled scientists and thinkers for decades. Named after physicist Enrico Fermi, this paradox poses a fundamental question: If the universe is vast and teeming with potentially habitable planets, where are the extraterrestrial civilizations?

At the heart of the Fermi Paradox lies the apparent contradiction between the high probability of intelligent life existing elsewhere in the galaxy and the lack of evidence for, or contact with, such civilizations. In other words, given the sheer number of stars and planets in the Milky Way alone, it seems highly likely that other intelligent beings should exist. Yet, we have not received any confirmed signals or direct evidence of their existence.

The paradox becomes even more intriguing when we consider the age of the universe. With billions of years of cosmic history, one would expect that, if other civilizations had arisen, they would have had ample time to advance and expand, potentially becoming highly advanced spacefaring cultures.

Several potential solutions to the Fermi Paradox have been proposed. One possibility is that intelligent civilizations are rare, and we happen to exist in a cosmic "dead zone" where life is an exception rather than the rule. Alternatively, it's possible that advanced civilizations may exist but have not yet reached a level of technology or curiosity to make their presence known to us. They may communicate through means beyond our current understanding, or they might deliberately avoid contact with emerging civilizations.

The Fermi Paradox serves as a reminder of the profound mysteries that the cosmos holds. It challenges our assumptions about the prevalence of intelligent life and encourages us to continue exploring and searching for answers. As we delve deeper into the cosmos, the question of "Where Are They?" remains an enigmatic and thought-provoking puzzle that drives our quest for knowledge and understanding of the universe.

1.2.2. Humanity's Quest

Humanity's quest for answers about the existence of extraterrestrial life is a testament to our insatiable curiosity and the innate desire to explore the unknown. This quest is deeply rooted in our collective imagination and has been a driving force throughout history.

From ancient civilizations peering up at the night sky to modern space exploration missions, the search for signs of life beyond Earth has been a consistent theme. It reflects our fundamental need to understand our place in the universe and to seek out companionship or knowledge beyond our home planet.

The quest for extraterrestrial life takes various forms, from the scientific endeavors of astronomers and astrobiologists to the imaginative narratives of science fiction writers and filmmakers. It has inspired the creation of powerful telescopes, space probes, and observatories, all aimed at unraveling the mysteries of the cosmos.

One of the driving factors behind this quest is the belief that the discovery of extraterrestrial life would be a profound moment in human history. It would not only expand our understanding of biology and evolution but also challenge our perspectives on the uniqueness of Earth and the diversity of life in the universe.

Moreover, the search for extraterrestrial life has practical implications. It informs our understanding of habitable zones in the universe, where conditions might be suitable for life to thrive. This knowledge could guide future space exploration missions, helping us identify potential destinations for further investigation.

As we continue our journey into the cosmos, humanity's quest for answers regarding the existence of extraterrestrial life remains a testament to our boundless curiosity, our thirst for knowledge, and our enduring belief that the universe holds untold secrets waiting to be uncovered. It is a quest that unites scientists, dreamers, and explorers in the pursuit of one of the most profound questions of our time.

1.2.3. Loneliness and Infinity

Loneliness and infinity may seem like two contrasting concepts, yet they are intricately intertwined when contemplating our place in the vast cosmos. As we gaze upon the stars and contemplate the sheer scale of the universe, feelings of both wonder and isolation often emerge.

The feeling of loneliness arises from the realization that our home, Earth, is just a tiny speck in the grand cosmic scheme. Our planet orbits an ordinary star in the outskirts of an average galaxy, and there are countless galaxies beyond our own, each containing billions of stars and potentially even more planets. In this context, our existence can feel insignificant and isolated.

Infinity, on the other hand, is a concept that stretches our comprehension to its limits. The universe is thought to be boundless, extending far beyond the reaches of our most powerful telescopes. The idea of infinity in space and time challenges our understanding and invites us to ponder questions that transcend our earthly experiences.

The juxtaposition of loneliness and infinity drives us to seek answers. We yearn to know if we are truly alone in the universe or if there are other intelligent beings out there, sharing in the awe of the cosmos. This quest for connection, both on a cosmic and existential level, fuels our exploration and our desire to reach out into the unknown.

The search for extraterrestrial life and the exploration of distant celestial objects are not only scientific endeavors but also a reflection of our human spirit. They offer a sense of purpose in our quest to overcome the loneliness that can accompany the contemplation of infinity.

While the universe's vastness and the potential solitude it implies can be daunting, they also inspire us to push the boundaries of knowledge and expand our horizons. Loneliness and infinity, when viewed through the lens of exploration and discovery, become driving forces that motivate us to reach for the stars and uncover the secrets of the universe.

CHAPTER 2

The Infinite Size of the Universe

2.1. The Formation and Expansion of the Universe

2.1.1. The Big Bang and the Birth of the Universe

The Big Bang theory stands as the cornerstone of our understanding of the universe's origin. It describes a momentous event that marked the birth of the cosmos as we know it. This theory proposes that the universe began as an incredibly hot and dense point, often referred to as a singularity, around 13.8 billion years ago.

At this primordial moment, the universe was a place of unimaginable conditions, where the laws of physics as we understand them today did not apply. Time and space as we know them were intricately connected and originated from this singularity. The term "Big Bang" refers to the rapid expansion and cooling of the universe that followed this initial moment.

As the universe expanded, it underwent a series of transformations. In its earliest moments, it was too hot for atoms to form, consisting instead of a dense plasma of particles, primarily protons, electrons, and photons. Over time, as it expanded and cooled, the universe allowed the formation of atomic nuclei and, eventually, atoms.

The release of cosmic microwave background radiation, a remnant from the early universe, provides compelling evidence supporting the Big Bang theory. This radiation, discovered in the mid-20th century, serves as a snapshot of the universe's state when it transitioned from a hot, dense state to one where matter could form.

The Big Bang theory not only explains the origin of the universe but also offers insights into its ongoing expansion. Observations have shown that galaxies are moving away from each other, suggesting that the universe is still expanding. This expansion has led to a universe filled with galaxies, stars, planets, and the potential for life.

Understanding the Big Bang and the universe's birth is not only a scientific endeavor but also a profound philosophical and existential exploration. It invites us to contemplate the ultimate questions of existence, the nature of time, and the origins of the cosmos itself. As we delve into the details of this theory, we uncover the remarkable story of how the universe, with all its complexity and diversity, began with a single explosive moment.

2.1.2. The Dance of Galaxies

The cosmos is a grand stage where galaxies perform an intricate dance, driven by the forces of gravity and cosmic dynamics. Understanding this cosmic ballet is essential to comprehending the structure and evolution of our universe.

Galaxies, vast collections of stars, gas, dust, and dark matter, come in various shapes and sizes. They can be spiral, elliptical, or irregular, each with its unique characteristics and history. These galaxies are not static; they are in constant motion, influenced by the gravitational pull of neighboring galaxies.

One of the defining features of galaxies is their organization into groups and clusters. Galaxies rarely exist in isolation; instead, they gather into cosmic communities. These clusters can contain just a handful of galaxies or thousands, bound together by gravity.

The dance of galaxies within these clusters is a complex interplay of attraction and repulsion. Gravity pulls galaxies toward one another, causing them to cluster together. Yet, the expansion of the universe exerts a counteracting force, causing galaxies to move apart. This delicate balance shapes the structure of the universe on a grand scale.

One of the most captivating aspects of this cosmic dance is the formation of galaxy filaments and voids. Galaxies tend to align along cosmic filaments—vast, thread-like structures that stretch across the universe. These filaments act as cosmic highways, guiding galaxies along their paths through space.

Galaxy collisions and mergers are another fascinating element of this dance. When galaxies come too close, their gravitational interaction can lead to dramatic events. Stars and gas clouds can be torn away from their home galaxies, while the merging galaxies can ultimately form new, larger galaxies.

As we unravel the complexities of the dance of galaxies, we gain insights into the formation and evolution of the universe itself. The study of galaxy dynamics and cosmic structures allows us to trace the history of the cosmos, from its earliest moments after the Big Bang to the present day.

This dance of galaxies is a testament to the elegance and intricacy of the universe's design. It reminds us that the cosmos is not a static backdrop but a dynamic, ever-changing stage where galaxies perform their celestial ballet, leaving their mark on the grand tapestry of space and time.

2.1.3. The Concept of Infinity and Its Boundaries

The concept of infinity is a profound and enigmatic aspect of the universe that challenges our understanding of both mathematics and the cosmos. Infinity represents the limitless, the unbounded, and the immeasurable, and it plays a crucial role in our exploration of the universe's vastness.

In mathematics, infinity is a symbol used to represent a quantity that has no upper bound. It defies finite measurement and encompasses the idea of endlessness. Mathematicians often use the symbol ∞ to denote infinity, and it is a fundamental concept in calculus, set theory, and other branches of mathematics.

Infinity also finds its place in cosmology and astrophysics, where it is intimately tied to the concept of the universe's size and age. The universe, as we understand it, is vast and seemingly boundless, stretching beyond the reaches of our most advanced telescopes.

Yet, while infinity represents the concept of limitless expanse, the universe may have boundaries, and this paradoxical interplay between the infinite and the finite is a topic of ongoing exploration. The nature of the universe's boundaries, if they exist, remains a subject of great debate among cosmologists.

One hypothesis suggests that the universe is finite but unbounded, similar to the surface of a sphere. In such a scenario, traveling in one direction for a sufficiently long time would eventually bring you back to your starting point. This concept challenges our intuitive understanding of space but is consistent with certain mathematical models of the universe.

Another perspective posits that the universe may indeed be infinite, extending infinitely in all directions. This notion implies that the cosmos has no inherent boundaries and goes on forever, a concept that can be challenging to grasp.

The concept of infinity and its boundaries prompt us to contemplate the nature of space, time, and the universe itself. It invites us to explore the limits of our knowledge and the boundaries of human understanding. Whether the universe is ultimately finite, infinite, or something else entirely, the concept of infinity continues to be a driving force behind our quest to unravel the mysteries of the cosmos.

2.2. The Number of Stars and Galaxies

2.2.1. The Milky Way and Earth's Place

The Milky Way, a sprawling and majestic spiral galaxy, serves as both our cosmic home and a profound point of reference in our exploration of the universe. Understanding the Milky Way's structure and our place within it is essential to comprehending our cosmic surroundings.

Our galaxy, the Milky Way, stretches across the night sky as a band of faint, milky light. It is a vast collection of stars, dust, gas, and dark matter, forming a magnificent spiral structure. Our solar system, including Earth, resides within one of the Milky Way's spiral arms, known as the Orion Arm or Local Spur.

Realizing our location within the Milky Way is a humbling experience. The Milky Way is a colossal galaxy, estimated to contain hundreds of billions of stars. Earth, situated in one of its spiral arms, is just one of countless planets orbiting one of these stars.

This galaxy is our celestial neighborhood, and it serves as a backdrop to our nightly view of the cosmos. It is a vast realm, where stars are born, evolve, and eventually fade away. Understanding our place within the Milky Way inspires a sense of connection to the greater universe and fosters a deeper appreciation for the night sky.

The Milky Way also serves as a reference point in our quest to locate other galaxies in the universe. When astronomers peer into the depths of space, they often use the Milky Way's stars and constellations as familiar markers to navigate the cosmos.

Moreover, studying the Milky Way provides insights into the nature of galaxies in general. It offers a unique perspective on galactic structures, star formation processes, and the intricate dance of stars within spiral arms.

As we contemplate the Milky Way and Earth's place within it, we are reminded of the vastness of the universe and the remarkable beauty of our home galaxy. It is a reminder that, in our quest for understanding the cosmos, we are but one small part of a much grander cosmic tapestry.

2.2.2. How Many Stars Exist in the Universe?

The question of how many stars exist in the universe is a profound and awe-inspiring inquiry that challenges our ability to grasp the vastness of cosmic scales. While it is impossible to provide an exact count, astronomers have developed estimates that shed light on the staggering number of stars scattered throughout the cosmos.

In our Milky Way galaxy alone, there are estimated to be between 100 billion and 400 billion stars. Each of these stars, like our Sun, is a luminous sphere of gas, radiating light and heat into space. Some stars are larger and more massive, while others are smaller and fainter, but collectively, they form the brilliant tapestry of our galaxy.

Beyond the Milky Way, the observable universe contains an estimated 2 trillion galaxies, each with its own complement of stars. This astonishing number is a testament to the vastness of space and the multitude of cosmic structures scattered throughout the universe.

To put this into perspective, if we consider just one grain of sand to represent a star, the number of stars in the observable universe would exceed the total number of grains of sand on all the beaches and deserts on Earth. It is a figure that stretches the limits of human imagination.

The quest to count the stars in the universe is a continuous effort. Astronomers use various methods, such as statistical analyses, observations of galaxy populations, and simulations, to arrive at these estimates. However, it is essential to acknowledge that these numbers are approximations, and the true number of stars in the universe may remain forever beyond our precise determination.

Nevertheless, contemplating the enormity of the universe and the countless stars it contains evokes a sense of wonder and humility. It reinforces our understanding that we are part of a cosmic panorama that extends far beyond the boundaries of our home planet. The stars in the universe are not just points of light in the night sky; they are the building blocks of galaxies, the crucibles of creation, and a source of endless fascination for astronomers and stargazers alike.

2.2.3. The Diversity and Significance of Stars

Stars, those radiant celestial bodies that illuminate the night sky, come in an astonishing array of sizes, colors, and behaviors. Their diversity and significance in the cosmos are fundamental to our understanding of the universe's structure and evolution.

Stars are born from vast clouds of gas and dust scattered throughout galaxies. The process of star formation begins when these cosmic nurseries undergo gravitational collapse. As the gas and dust condense under their own gravity, they heat up and ignite nuclear fusion in their cores, marking the birth of a new star.

The size of a star plays a crucial role in determining its fate. Stars come in various sizes, ranging from diminutive red dwarfs, which are only a fraction of the Sun's mass, to massive supergiants, which can be tens or even hundreds of times more massive than our Sun.

The color of a star is closely linked to its temperature. Hotter stars appear bluish-white, while cooler ones emit a reddish hue. This color diversity is a testament to the wide range of temperatures and physical conditions within the universe's stellar population.

Stars are not static objects; they undergo a life cycle characterized by stages such as birth, middle age, and eventual demise. Some stars, like our Sun, follow a relatively stable path and will eventually transition into a red giant before settling into a white dwarf stage. Others, much more massive, may end their lives in spectacular supernova explosions, leaving behind remnants like neutron stars or black holes.

The significance of stars goes beyond their role as cosmic beacons of light. They are the engines of the universe, producing the elements that make up planets, moons, and life itself. The fusion reactions in a star's core forge heavier elements, which are eventually released into space when the star reaches the end of its life cycle. These elements become the building blocks for new generations of stars and planetary systems.

Moreover, stars serve as celestial laboratories, allowing astronomers to study the fundamental forces and processes that govern the universe. By analyzing the light emitted by stars, scientists can decipher their compositions, temperatures, and distances, providing insights into the cosmos's structure and history.

In summary, stars are not only captivating objects of wonder in the night sky but also essential components of the cosmic narrative. Their diversity and significance illuminate the rich tapestry of the universe, shaping its past, present, and future, and inviting us to explore the mysteries of the cosmos.

2.3. Possibilities of Extraterrestrial Life

2.3.1. Habitability and the Goldilocks Zone

Habitability, the concept of conditions suitable for life, is a central theme in our exploration of the cosmos. Understanding where and how life can thrive in the universe leads us to the concept of the Goldilocks Zone, a critical consideration in the search for potentially habitable environments beyond Earth.

The Goldilocks Zone, also known as the habitable zone or the circumstellar habitable zone, refers to the region around a star where conditions are just right for liquid water to exist on the surface of a planet. Liquid water is a fundamental requirement for life as we know it, making this zone a crucial factor in assessing a celestial body's potential for habitability.

The Goldilocks Zone is defined by a delicate balance between a star's radiation and a planet's distance from that star. If a planet orbits too close to its star, it becomes too hot, and water would exist primarily as vapor or be lost to space. Conversely, if a planet orbits too far from its star, it becomes too cold, and water freezes, rendering the surface inhospitable.

Within this zone, conditions are "just right," akin to the porridge in the story of Goldilocks and the Three Bears—not too hot and not too cold. Planets within the Goldilocks Zone have the potential to maintain stable surface temperatures, allowing liquid water to exist, a key ingredient for life as we understand it.

The search for exoplanets (planets outside our solar system) within the Goldilocks Zone has become a focal point in the quest to find habitable environments and perhaps even extraterrestrial life. Astronomers employ various techniques, such as transit observations and radial velocity measurements, to identify exoplanets situated within this critical region around other stars.

While the Goldilocks Zone is a fundamental concept, it is not the sole determinant of habitability. Factors such as a planet's atmosphere, geological activity, magnetic field, and many others also influence its potential to support life.

The study of habitability and the Goldilocks Zone not only guides our search for habitable worlds but also deepens our appreciation for the delicate balance of conditions that make Earth a haven for life. It underscores the uniqueness of our planet while inspiring the exploration of distant celestial bodies where life, in some form, might exist, awaiting discovery in the vast cosmic tapestry.

2.3.2. Candidates for Exoplanets

In the quest to discover exoplanets—planets located beyond our solar system—a wide array of celestial objects has emerged as promising candidates. Identifying these candidates involves a combination of astronomical techniques and instruments designed to detect the subtle signs of these distant worlds.

Transiting Exoplanets: Many exoplanets are discovered through the transit method. When a planet passes in front of its host star from our vantage point, it causes a temporary dip in the star's brightness. This periodic dimming of the star's light, known as a transit, allows astronomers to infer the presence and characteristics of the orbiting exoplanet. Thousands of exoplanets have been discovered using this technique.

Radial Velocity Candidates: Some exoplanets reveal themselves through the radial velocity method. As a planet orbits a star, its gravitational pull causes the star to wobble slightly. This motion leads to observable shifts in the star's spectral lines due to the Doppler effect. By carefully analyzing these shifts, astronomers can deduce the presence and properties of the orbiting exoplanet.

Direct Imaging Prospects: Direct imaging involves capturing actual images of exoplanets. This technique is challenging due to the vast contrast in brightness between a star and its planets. However, advanced instruments and adaptive optics systems are enabling the direct detection of exoplanets, particularly those orbiting farther from their host stars.

Microlensing Events: Gravitational microlensing occurs when the gravitational field of a massive object, like a star, bends and amplifies the light of a more distant star. If a planet orbits the foreground star, it can create additional microlensing effects. Detecting these transient events helps identify exoplanet candidates.

Astrometry and Astroseismology: Astrometry involves measuring the precise positions of stars over time.

When a star wobbles due to an orbiting planet, astrometry can reveal the planet's presence. Astroseismology, on the other hand, studies star oscillations caused by exoplanet interactions, providing clues about the planet's characteristics.

Habitable Zone Candidates: Special attention is given to exoplanets within the habitable zone or Goldilocks Zone of their host stars. These planets, where conditions might support liquid water, are prime targets in the search for potentially habitable environments and extraterrestrial life.

Exomoon Possibilities: In recent years, astronomers have also explored the detection of exomoons—moons orbiting exoplanets. The study of these celestial duos adds another layer of complexity to the search for habitable environments beyond our solar system.

The search for exoplanets is a dynamic field, driven by technological advancements and a growing database of candidates. While many candidates have been identified, the confirmation and characterization of these distant worlds remain ongoing challenges. Nevertheless, each candidate brings us closer to unraveling the cosmic diversity of planets and the potential for life beyond Earth.

2.3.3. Chemistry and the Origin of Life

The quest to understand the origin of life is a fundamental pursuit in the field of astrobiology. This quest delves deep into the realms of chemistry, exploring the processes and conditions that may have given rise to life on Earth and, potentially, on other celestial bodies in the universe.

Chemistry serves as the bridge between the inanimate and the living, providing insights into how the building blocks of life, such as amino acids, nucleotides, and lipids, can form and interact. Understanding these chemical processes sheds light on the plausibility of life emerging elsewhere in the cosmos.

One of the key questions in the study of life's origin revolves around the nature of the primordial soup—the mixture of chemicals and compounds that existed on Earth's early surface. It is believed that the conditions on the young Earth, with its abundant supply of water, minerals, and energy sources, were conducive to the formation of complex organic molecules.

Experiments simulating these conditions have demonstrated that simple organic molecules, including amino acids and nucleotides, can arise spontaneously through chemical reactions. These molecules are the building blocks of proteins and nucleic acids, the essential components of living organisms.

Furthermore, hydrothermal vents on the ocean floor have been proposed as potential cradles of life. These extreme environments, rich in minerals and hot water, provide a unique setting where chemical reactions can occur, leading to the formation of organic molecules.

The chemistry of life is intimately tied to the concept of chirality—the handedness of molecules. Living organisms on Earth predominantly use left-handed amino acids and right-handed sugars. The origin of this homochirality is a topic of ongoing research and has implications for understanding the emergence of life's biochemical processes.

As we explore the chemistry of life's origins, we gain insights not only into the past but also into the potential for life elsewhere. The discovery of complex organic molecules on celestial bodies like comets and asteroids hints at the possibility of life's precursors existing beyond Earth.

The study of chemistry and the origin of life is a multidisciplinary endeavor that combines elements of biology, chemistry, geology, and astrophysics. It invites us to ponder the cosmic connections between the chemistry of the universe and the remarkable emergence of life on our own planet, as well as the potential for life to arise elsewhere in the vast expanse of the cosmos.

2.4. Mysteries of the Universe and Human Curiosity

2.4.1. The Allure of the Uncharted

The allure of the uncharted has always been a driving force in our exploration of the cosmos. It represents the innate human desire to push the boundaries of knowledge and venture into the unknown, whether by sending spacecraft to distant planets or peering into the depths of the universe with powerful telescopes.

One of the most captivating aspects of the uncharted is the prospect of discovering new worlds. From the exploration of our own solar system to the search for exoplanets around distant stars, the idea of finding new celestial bodies ignites our curiosity and fuels our quest for understanding.

The uncharted also extends to the mysteries of the cosmos, from the enigmatic nature of dark matter and dark energy to the perplexing behavior of black holes. These cosmic puzzles challenge our understanding of the fundamental laws of the universe and beckon us to seek answers beyond our current knowledge.

Space exploration is a prime example of humanity's fascination with the uncharted. Sending spacecraft to other planets, moons, and asteroids allows us to collect data and images from previously unexplored realms. These missions not only expand our scientific understanding but also inspire wonder and awe as we gaze upon landscapes and features that no human has seen before.

The uncharted is not limited to our solar system. Telescopes like the Hubble Space Telescope and its successors enable us to peer deep into the universe's past, capturing light that has traveled billions of years to reach us. These observations reveal distant galaxies, cosmic phenomena, and the cosmic web of structure that stretches across the cosmos.

Moreover, the allure of the uncharted extends to the search for extraterrestrial life. Whether through the study of extremophiles on Earth or the exploration of potentially habitable environments on other celestial bodies, scientists are driven by the tantalizing possibility of discovering life beyond our planet.

As we continue to explore the uncharted, we are propelled by the human spirit of curiosity and the belief that the cosmos holds untold secrets waiting to be uncovered. It is this relentless drive to venture into the unknown that propels us into the vast and mysterious universe that surrounds us, reminding us of the limitless possibilities that await our discovery.

2.4.2. Humanity's Gaze into Space

Throughout history, humanity's gaze into space has been marked by wonder, curiosity, and a relentless pursuit of understanding the cosmos. This enduring fascination with the celestial realms has driven us to explore, observe, and contemplate the mysteries of the universe.

The act of looking up at the night sky has been a common human experience for millennia. Ancient civilizations observed the movements of stars and planets, weaving intricate stories and mythologies to explain their celestial dance. These early stargazers laid the foundation for our evolving relationship with the cosmos.

The invention of telescopes in the early modern era revolutionized our ability to peer deeper into space. Astronomers like Galileo Galilei and Johannes Kepler made groundbreaking discoveries, from the phases of Venus to the laws of planetary motion, reshaping our understanding of the solar system and our place within it.

The exploration of space, marked by missions like Apollo 11's historic moon landing, expanded our horizons beyond Earth's boundaries. Humans left footprints on another world, capturing the imagination of people around the globe and demonstrating our capacity to overcome seemingly insurmountable challenges.

Today, observatories and space telescopes continue to provide us with breathtaking views of distant galaxies, nebulae, and cosmic phenomena. These images not only inspire a sense of wonder but also fuel scientific inquiry, helping us unravel the secrets of the universe's origins and evolution.

Moreover, our gaze into space extends beyond the visual spectrum. Telescopes and instruments detect a wide range of electromagnetic radiation, from radio waves to X-rays and gamma rays. This multisensory approach to cosmic exploration allows us to perceive celestial objects and phenomena that are invisible to the naked eye.

The search for exoplanets, habitable environments, and signs of extraterrestrial life represents a modern frontier in our quest to understand the cosmos. We eagerly scan the night sky and analyze vast datasets, hoping to find clues to the existence of life beyond Earth.

In essence, humanity's gaze into space is not merely a scientific endeavor but also a deeply philosophical and existential one. It prompts us to ponder the nature of the universe, our place within it, and the profound questions that lie at the intersection of science and wonder.

As we continue to look up at the stars and explore the uncharted regions of the cosmos, we honor our human heritage of curiosity and exploration. The act of gazing into space connects us to the countless generations that have come before us and to the boundless possibilities that await our future discoveries.

2.4.3. First Steps Toward Exploration

The journey of cosmic exploration has been shaped by a series of pivotal first steps that mark significant milestones in our quest to understand the universe. These early forays into space and beyond have paved the way for humanity's continued exploration of the cosmos.

The Space Age Dawns: The launch of the Soviet satellite Sputnik 1 on October 4, 1957, marked the beginning of the Space Age. This tiny orbiter, equipped with a radio transmitter, became the first human-made object to orbit Earth, ushering in an era of space exploration.

The First Human in Space: On April 12, 1961, Yuri Gagarin, a Soviet cosmonaut, became the first human to journey into space aboard Vostok 1. His historic orbit around Earth lasted just 108 minutes but opened the door to human spaceflight.

Landing on the Moon: Apollo 11, with astronauts Neil Armstrong and Buzz Aldrin, achieved humanity's first lunar landing on July 20, 1969. Neil Armstrong's iconic words as he descended onto the lunar surface, "That's one small step for [a] man, one giant leap for mankind," echoed throughout history.

Exploring the Red Planet: The Mars Exploration Rover mission, with the twin rovers Spirit and Opportunity, landed on Mars in 2004. These robotic explorers provided valuable insights into the Martian surface and confirmed the presence of water in the planet's past.

Voyaging Beyond the Solar System: Launched in 1977, the Voyager 1 and Voyager 2 spacecraft embarked on a journey beyond our solar system. Voyager 1 became the first human-made object to enter interstellar space in 2012, carrying a golden record with greetings from Earth.

Observing the Cosmos: Telescopes like the Hubble Space Telescope, launched in 1990, have revolutionized our understanding of the universe. Hubble's stunning images and observations have deepened our knowledge of distant galaxies, nebulae, and cosmic phenomena.

International Cooperation: The International Space Station (ISS) exemplifies international collaboration in space exploration. This orbiting laboratory has hosted astronauts from multiple countries, fostering scientific research and cooperation in space.

These first steps toward exploration represent humanity's collective determination to venture into the unknown and unravel the mysteries of the cosmos. They highlight the significance of cooperation among nations, the spirit of discovery, and the enduring human curiosity that continues to drive us further into the uncharted realms of space.

CHAPTER 3

Discovery of Exoplanets

3.1. Kepler and Exoplanets Beyond Earth

3.1.1. The Kepler Space Telescope and Its Discoveries

The Kepler Space Telescope stands as one of the most significant instruments in the quest to unveil the secrets of our galaxy and beyond. Its mission, dedicated to the search for exoplanets, led to a revolution in our understanding of the prevalence and diversity of planets beyond our solar system.

Kepler's Vision: Launched on March 7, 2009, Kepler was specifically designed to detect exoplanets by observing tiny, periodic dips in the brightness of stars. These dips, known as transits, occur when an exoplanet passes in front of its host star from Earth's perspective, causing a slight decrease in the star's brightness.

A Treasure Trove of Exoplanets: Kepler's primary mission lasted until 2013, during which it monitored over 150,000 stars in a patch of the Milky Way. It discovered thousands of exoplanet candidates, ranging from small, rocky worlds to gas giants. Some of these planets orbit within their stars' habitable zones, igniting hopes of finding environments suitable for life.

Kepler's Exoplanet Catalog: The telescope's data led to the creation of the Kepler Exoplanet Catalog, a repository of confirmed exoplanet discoveries. Kepler's meticulous observations allowed scientists to determine the size, orbital characteristics, and potential habitability of these distant worlds.

The Kepler-90 System: One of Kepler's notable discoveries was the Kepler-90 system, which contains eight planets orbiting a Sun-like star. This system demonstrated the richness of multi-planet systems and expanded our knowledge of planetary architectures.

Legacy of Exploration: While Kepler's primary mission concluded, its legacy continues through extended missions and the wealth of data it generated. The Transiting Exoplanet Survey Satellite (TESS), launched in

2018, follows in Kepler's footsteps, searching for exoplanets around nearby stars.

The Kepler Space Telescope's contributions to the field of exoplanet science have been immeasurable. It redefined our understanding of the universe by revealing the prevalence of exoplanets and inspiring further exploration into the mysteries of these distant worlds. Kepler's legacy persists as humanity's gaze remains fixed on the stars, ever in search of new horizons in the cosmic journey.

3.1.2. Diversity of Exoplanets

The exploration of exoplanets, planets located outside our solar system, has unveiled a remarkable tapestry of diversity in the cosmos. Exoplanets come in an astonishing array of sizes, compositions, and orbital configurations, challenging our preconceptions about planetary systems.

Variety of Sizes: Exoplanets exhibit a wide range of sizes, from diminutive worlds, only a few times larger than Earth, to massive gas giants, many times the size of

Jupiter. These size variations hint at the vast spectrum of planetary formations that occur throughout the universe.

Rocky Exoplanets: Rocky or terrestrial exoplanets, akin to Earth, are of particular interest. These worlds are composed primarily of solid material, such as rock and metal. The search for rocky exoplanets within the habitable zones of their stars represents a focal point in the quest for potentially habitable environments.

Gas Giants: Gas giants, like Jupiter and Saturn, are common inhabitants of other star systems. These massive planets often reside farther from their host stars, showcasing the diversity of planetary architectures. Some exoplanetary systems even contain multiple gas giants, each with its unique orbit.

Exoplanet Atmospheres: Observations of exoplanet atmospheres have revealed a fascinating array of compositions. Some exoplanets have atmospheres rich in hydrogen and helium, akin to our gas giants, while others exhibit diverse chemistry, including clouds of water vapor, carbon dioxide, or methane.

Exoplanetary Moons: Moons orbiting exoplanets add another layer of complexity to these systems. Just as

Earth's moon plays a vital role in our planet's dynamics, exomoons can influence their parent planets' environments and potential habitability.

Exoplanets in Binary Systems: In some cases, exoplanets orbit binary star systems, where two stars are gravitationally bound. These complex environments challenge our understanding of planetary dynamics and habitability factors.

Eccentric Orbits: The orbital configurations of exoplanets can be highly eccentric, with elliptical paths that vary significantly in distance from their host stars. Such eccentricities can lead to extreme temperature variations and unique planetary climates.

Exoplanetary Diversity and Habitability: The diversity of exoplanets underscores the complexity of planetary systems in the universe. It inspires scientists to consider the potential for habitability in a wide range of environments, from scorching gas giants to frigid ice worlds.

In our exploration of exoplanets, we encounter a vast mosaic of celestial bodies that expand our horizons and deepen our understanding of planetary systems beyond our solar system. Each new discovery adds a brushstroke to this cosmic portrait, emphasizing the boundless variety of worlds awaiting our exploration in the grand tapestry of the universe.

3.1.3. Orbits and Characteristics of Planets

The study of exoplanets involves a detailed examination of their orbits and characteristics, offering valuable insights into the diversity and behavior of these distant worlds as they orbit their host stars.

Eccentric and Circular Orbits: Exoplanets exhibit a wide range of orbital shapes, from nearly circular paths to highly eccentric, elongated orbits. The eccentricity of an orbit affects the planet's distance from its host star, leading to temperature variations and climatic differences.

Orbital Periods: The time it takes for an exoplanet to complete one orbit around its star, known as its orbital period, can vary significantly. Some planets zip around their stars in just a few Earth days, while others have orbits that extend for several Earth years.

Close-In Hot Jupiters: A fascinating subset of exoplanets includes "hot Jupiters," massive gas giants that orbit incredibly close to their host stars. These planets experience scorching temperatures, challenging our understanding of planetary formation and migration.

Habitability Factors: The distance between an exoplanet and its host star, known as the semi-major axis, plays a crucial role in determining a planet's potential habitability. Planets within the habitable zone, where conditions for liquid water are favorable, are of particular interest in the search for life beyond Earth.

Atmospheric Composition: Observations of exoplanet atmospheres provide essential clues about their compositions. Some exoplanets have atmospheres dominated by hydrogen and helium, while others feature a diverse mix of gases, including water vapor, carbon dioxide, and methane.

Transit Method Discoveries: The transit method, which detects exoplanets by observing the slight dimming of a star's brightness as a planet passes in front of it, has yielded a wealth of data. This method allows scientists to measure a planet's size, orbital period, and potential habitability.

Exoplanetary Systems: Many exoplanets are part of multi-planet systems, akin to our solar system. These systems showcase the intricate dynamics of planets as they interact with one another and their host stars.

Exoplanetary Characteristics: The study of exoplanetary characteristics extends beyond size and orbital parameters. It includes investigations into planetary atmospheres, magnetic fields, surface conditions, and the potential for moons or rings.

Search for Exomoons: Researchers also explore the possibility of exomoons—moons orbiting exoplanets. The presence of moons can influence a planet's habitability and add complexity to these distant worlds.

The analysis of orbits and characteristics of exoplanets offers a comprehensive view of the cosmos's planetary diversity. These discoveries challenge our preconceptions about planetary systems and provide a foundation for exploring the potential for life beyond Earth. Each new observation adds a piece to the puzzle, bringing us closer to understanding the complex tapestry of exoplanetary worlds.

3.2. Habitability and the Goldilocks Zone

3.2.1. Conditions for Life

The search for extraterrestrial life hinges on identifying environments and conditions conducive to life as we understand it. Exploring the fundamental prerequisites for life broadens our comprehension of where and how life could potentially exist beyond the confines of Earth.

One of the most critical factors in the search for life is the presence of liquid water. Liquid water is a universal solvent and a vital medium for the biochemical reactions that sustain life on Earth. Its existence, whether on the surface of a celestial body or beneath layers of ice, serves as a pivotal indicator of potential habitability.

Another essential condition for life, as we know it, is a stable temperature range that supports the fundamental processes of biology. While extremophiles demonstrate life's adaptability to extreme conditions, moderate temperatures within a habitable zone remain a primary consideration in our exploration of potentially habitable environments.

Life also relies on specific chemical building blocks, including carbon, hydrogen, oxygen, nitrogen, phosphorus, and sulfur (CHONPS). These elements form the foundation of organic molecules like amino acids and nucleotides, which are essential for the intricate processes of life.

Furthermore, all known life on Earth depends on a reliable source of energy, often derived from a star, like our Sun, to power biological functions. Alternative energy sources, such as geothermal or chemical energy, can also support life in extreme environments.

Protecting life from harmful cosmic and solar radiation is of paramount importance for its survival. Various mechanisms, such as planetary atmospheres, magnetic fields, or subterranean environments, can provide the necessary shielding against detrimental radiation.

Essential nutrients, including vital minerals and trace elements, are necessary for life's metabolic processes. The availability and cycling of these nutrients contribute to the potential habitability of celestial bodies.

A stable environment, characterized by relatively consistent conditions over geological timescales, provides the foundation for the development and persistence of life. Drastic environmental fluctuations can pose challenges to the continuity of life forms.

Water's unique properties as a solvent are central to life's chemical processes. Its ability to dissolve and transport ions and molecules plays a fundamental role in the biochemical reactions that underpin life.

The pH level, reflecting the acid-base balance, is a crucial factor for life on Earth. Maintaining suitable pH levels is vital for enzyme activity and various cellular processes.

Finally, the environmental sustainability of habitable regions, including their ability to support food webs and ecosystems, plays a significant role in assessing the potential for life to thrive beyond our home planet.

3.2.2. The Significance of Habitable Zones

Habitable zones, often referred to as "Goldilocks zones," play a pivotal role in the search for potential life beyond Earth. These regions around stars represent the sweet spot where conditions may be just right to support the existence of liquid water and, by extension, life as we understand it.

Within a solar system, the habitable zone is defined as the region where a planet can maintain the necessary conditions for liquid water to exist on its surface. This concept is grounded in the understanding that water is a fundamental ingredient for life, and its presence significantly increases the likelihood of habitability.

The location of a habitable zone is intimately tied to the characteristics of its parent star. Stars come in various sizes and temperatures, and these factors determine the position and breadth of their habitable zones. Smaller, cooler stars have habitable zones closer in, while larger, hotter stars have habitable zones farther out.

The significance of habitable zones lies in their potential to harbor Earth-like conditions. Planets within these zones have the opportunity to maintain stable temperatures suitable for liquid water, a critical component for life's processes. However, habitable zones are not the sole determinants of a celestial body's habitability. Other factors, such as atmospheric composition, geology, and magnetic fields, also contribute to a planet's ability to sustain life.

The study of habitable zones extends beyond our solar system. Astronomers search for exoplanets located within the habitable zones of distant stars, using a variety of detection methods. The discovery of exoplanets in these zones raises tantalizing questions about the potential for life beyond our solar system and underscores the significance of these regions in our quest to unravel the mysteries of the universe.

In summary, habitable zones are critical in the search for extraterrestrial life due to their potential to host conditions conducive to life as we know it. These regions serve as a focal point in the exploration of celestial bodies and expand our understanding of where life may thrive beyond the confines of our home planet.

3.2.3. Defining Habitable Planets

The search for habitable planets represents a fundamental aspect of our quest to find life beyond Earth. Defining what makes a planet habitable involves a careful consideration of various factors that contribute to the potential for life to thrive in its unique environment.

Liquid Water: A primary criterion for habitability is the presence of liquid water. Water is an essential solvent for life's biochemical reactions, making it a crucial ingredient. The ability of a planet to maintain liquid water on its surface or subsurface significantly influences its habitability.

Stable Climate: Habitability depends on a stable and moderate climate. Extreme temperature fluctuations can be detrimental to life. A habitable planet should have mechanisms in place to regulate its climate and maintain a relatively consistent environment.

Atmosphere Composition: The composition of a planet's atmosphere plays a critical role in habitability. An atmosphere with the right balance of gases, including oxygen and nitrogen, is essential for supporting life as we know it. The presence of greenhouse gases can also help regulate temperatures.

Protection from Radiation: Protection from harmful cosmic and solar radiation is vital for habitability. Planets with magnetic fields, atmospheres, or geological features that shield against radiation are more likely to be habitable.

Nutrient Availability: Habitability is influenced by the availability of essential nutrients. These nutrients include minerals, trace elements, and organic compounds that support life's metabolic processes.

Energy Sources: Habitability often hinges on the availability of energy sources. Planets must receive an adequate amount of energy, typically from their host star, to sustain life processes. Alternative sources like geothermal or chemical energy can also contribute to habitability.

Geological Activity: Planets with geological activity, such as volcanism and tectonic processes, may be more habitable. These processes can recycle nutrients, maintain a stable environment, and provide energy sources.

Stable Orbit: A planet's orbit should be stable over long periods to maintain habitable conditions. Stability prevents extreme variations in temperature and ensures a consistent climate.

Moons and Tides: Moons can play a role in habitability by influencing a planet's tides. Tidal forces can mix nutrients in oceans, potentially fostering the development of life.

Ecosystem Potential: Habitability extends beyond the physical environment to include the potential for ecosystems to develop and thrive. Factors like food webs, biodiversity, and ecological niches contribute to a planet's habitability.

Defining habitable planets involves a comprehensive assessment of these factors, recognizing that habitability is a complex interplay of physical, chemical, and environmental conditions. As we continue our search for habitable worlds within and beyond our solar system, a deeper understanding of these criteria guides our exploration and informs our expectations for finding life elsewhere in the universe.

3.3. The Future of Exoplanet Research

3.3.1. Advances in Observation Technologies

The quest to explore the cosmos and seek answers to questions about extraterrestrial life has been greatly propelled by remarkable advances in observation technologies. These advancements have revolutionized our ability to study distant celestial bodies and uncover clues about the potential existence of life beyond Earth.

Telescopes: Telescopes have been the backbone of astronomical research for centuries. Modern telescopes equipped with cutting-edge optics and detectors allow scientists to observe distant stars, galaxies, and exoplanets with unprecedented clarity. Space-based observatories like the Hubble Space Telescope provide distortion-free views of the universe, enabling the study of exoplanetary atmospheres and distant galaxies.

Space Probes and Rovers: Robotic spacecraft, such as Mars rovers and interplanetary probes, have ventured to distant worlds within our solar system. These missions have provided valuable data on planetary geology, atmospheres, and the potential for past or present life on celestial bodies like Mars, Europa, and Enceladus.

Radio Telescopes: Radio telescopes have revolutionized our understanding of the universe by detecting radio waves emitted by celestial objects. These telescopes enable astronomers to study distant galaxies, pulsars, and cosmic microwave background radiation. The Search for Extraterrestrial Intelligence (SETI) employs radio telescopes to listen for potential extraterrestrial signals.

Exoplanet Detection Techniques: Advancements in exoplanet detection techniques, such as the transit method and radial velocity method, have led to the discovery of thousands of exoplanets. These techniques allow astronomers to identify planets in distant star systems, providing a wealth of data on their characteristics, orbits, and potential habitability.

Spectroscopy: Spectroscopy is a powerful tool for analyzing the composition of celestial objects. It allows scientists to determine the chemical makeup of stars, planets, and their atmospheres. Spectroscopic data is crucial for assessing the potential habitability of exoplanets.

Astrobiology Instruments: Space missions and rovers have been equipped with instruments designed specifically for astrobiology research. These tools can analyze soil samples, detect organic molecules, and search for signs of past or present microbial life.

Deep Space Telescopes: Upcoming space telescopes, like the James Webb Space Telescope (JWST), promise to revolutionize our understanding of the universe. JWST's advanced infrared capabilities will enable the study of distant galaxies, exoplanets, and the early universe.

Data Analysis and Artificial Intelligence: Advances in data analysis techniques, including artificial intelligence and machine learning, have become essential for processing the vast amounts of data collected from telescopes and spacecraft. These technologies help identify exoplanets, analyze spectra, and search for potential biosignatures.

Interferometry: Interferometers combine the signals from multiple telescopes to create detailed, high-resolution images of celestial objects. This technique has been used to study the surfaces of stars, image exoplanets, and investigate the centers of galaxies.

Planetary Protection: As missions explore potentially habitable worlds, planetary protection protocols and technologies have evolved to prevent contamination and protect against forward and backward contamination, ensuring the integrity of astrobiology research.

Advances in observation technologies continue to push the boundaries of our knowledge, fueling our curiosity about the cosmos, and enhancing our ability to search for signs of extraterrestrial life. These technological achievements underscore the importance of ongoing research and exploration efforts in our quest to understand the broader universe and our place within it.

3.3.2. Upcoming Exoplanet Discovery Missions

The field of exoplanetary science is on the cusp of a new era with several exciting missions on the horizon. These upcoming missions are poised to revolutionize our understanding of exoplanets, their diversity, and their potential for habitability.

James Webb Space Telescope (JWST): Scheduled for launch, the JWST is one of the most highly anticipated missions. Equipped with advanced infrared capabilities, this telescope will investigate the atmospheres of exoplanets in detail. It promises to uncover crucial information about their compositions, temperatures, and potential habitability.

PLATO (PLAnetary Transits and Oscillations of stars): PLATO, a European Space Agency mission, aims to detect and characterize exoplanets using the transit method. By observing the slight dimming of stars as planets pass in front of them, PLATO will identify and study a multitude of exoplanets, contributing to our understanding of their properties.

TESS (Transiting Exoplanet Survey Satellite): TESS, launched by NASA, is an ongoing mission that surveys the entire sky to identify exoplanets through transits. It focuses on nearby stars, providing valuable data on the sizes, orbits, and atmospheres of numerous exoplanets.

ARIEL (Atmospheric Remote-sensing Infrared Exoplanet Large-survey): ARIEL, an ESA mission, will delve into the atmospheres of exoplanets using spectroscopy in the infrared range. It aims to uncover key details about exoplanetary atmospheres and gain insights into their potential habitability.

CHEOPS (Characterizing Exoplanet Satellite): CHEOPS, an ESA mission, focuses on studying exoplanets already known to astronomers. It will precisely measure the sizes of these exoplanets, contributing to our understanding of their compositions and properties.

LUVOIR (Large UV/Optical/Infrared Surveyor): Proposed as a future NASA mission, LUVOIR envisions a powerful observatory capable of detecting and characterizing a wide range of exoplanets. Its capabilities may include direct imaging and spectroscopy of potentially habitable exoplanets.

HabEx (Habitable Exoplanet Observatory): Another NASA proposal, HabEx, is designed to search for habitable planets and signs of life. It would employ advanced imaging techniques to directly capture images of exoplanets and study their atmospheres.

WFIRST (Wide-Field Infrared Survey Telescope): While not exclusively an exoplanet mission, WFIRST will contribute significantly to exoplanet research by conducting surveys that may discover thousands of new exoplanets and study their atmospheres.

LISA (Laser Interferometer Space Antenna): LISA, a joint ESA-NASA mission, focuses on detecting gravitational waves. While its primary goal is not exoplanets, it may indirectly contribute by detecting the gravitational influence of massive exoplanets on their host stars.

BRITE-Constellation: The BRITE (BRIght Target Explorer) constellation consists of nanosatellites equipped with telescopes to observe bright stars. While not focused solely on exoplanets, it contributes to stellar research that informs our understanding of exoplanetary systems.

These upcoming exoplanet discovery missions promise to unlock new insights into the diversity of exoplanets, their atmospheres, and their potential habitability. They represent a significant step forward in our ongoing quest to uncover the secrets of the universe and the existence of life beyond Earth.

3.3.3. Searching for Traces of Extraterrestrial Life

One of the most profound questions in the field of astrobiology is the search for evidence of extraterrestrial life. Scientists employ various methods and missions to investigate celestial bodies and search for traces that could indicate the presence of life beyond Earth.

Mars Exploration: Mars has been a focal point in the search for extraterrestrial life. Rovers like Curiosity and Perseverance investigate the Martian surface for signs of past or present microbial life. The search includes examining ancient riverbeds, lakebeds, and subsurface environments.

Europa and Enceladus: Moons of Jupiter and Saturn, such as Europa and Enceladus, have subsurface oceans beneath icy crusts. Missions like NASA's Europa Clipper and the study of Enceladus by Cassini have identified plumes of water erupting from these moons, raising the possibility of subsurface life.

Sample Return Missions: Upcoming missions aim to return samples from celestial bodies, such as the Mars Sample Return mission and the OSIRIS-REx mission to the asteroid Bennu. These samples could contain potential traces of ancient life or organic molecules.

Search for Biosignatures: Scientists analyze the atmospheres of exoplanets for biosignatures—indicators that could suggest the presence of life. These biosignatures include oxygen, methane, and other gases that may be produced by living organisms.

Extremophiles on Earth: Studying extremophiles—microorganisms that thrive in extreme environments on Earth—provides insights into where and how life might exist beyond our planet. Extremophiles in acidic hot springs, deep-sea hydrothermal vents, and Antarctica's Dry Valleys inform astrobiological research.

SETI (Search for Extraterrestrial Intelligence): SETI researchers listen for potential signals from intelligent extraterrestrial civilizations using radio telescopes. The search includes examining candidate stars and analyzing radio emissions for patterns that may indicate intelligent communication.

Microbial Life in Extreme Environments: Researchers investigate extreme environments on Earth, such as hydrothermal vents and subglacial lakes, to understand how life can thrive in seemingly inhospitable conditions. These studies inform the search for life on other celestial bodies with extreme environments.

Astrobiology Instruments on Spacecraft: Space missions are equipped with instruments designed to search for signs of life. For example, the Mars Science Laboratory carries the SAM (Sample Analysis at Mars) instrument suite to analyze the composition of Martian soil and rock samples.

Future Missions to Ocean Worlds: Proposed missions to ocean worlds like Europa, Enceladus, and Titan aim to explore subsurface oceans for signs of life. These missions may involve drilling through ice or studying plumes for potential biomarkers.

Deep Space Telescopes: Upcoming telescopes, like the James Webb Space Telescope (JWST), will study the atmospheres of exoplanets, searching for biosignatures and potential habitable conditions.

The search for traces of extraterrestrial life is a multidisciplinary endeavor that spans various celestial bodies and scientific disciplines. It reflects humanity's deep-seated curiosity about our place in the universe and the possibility of life beyond our home planet.

3.4. The Enigma of Exoplanets

3.4.1. Could We Be Alone in the Universe?

The question of whether we are alone in the universe is one of the most profound and enduring mysteries that science and humanity as a whole have contemplated. Exploring this question requires a comprehensive examination of various factors and considerations that shape our understanding of the potential existence of extraterrestrial life.

The universe is incomprehensibly vast, with billions of galaxies, each containing billions of stars and likely an even greater number of planets. The sheer number of celestial bodies raises the statistical likelihood of other habitable worlds.

The discovery of thousands of exoplanets in the habitable zones of their stars suggests that planets like Earth may be common in the galaxy. This abundance of potentially habitable worlds intensifies the question of whether life could emerge elsewhere.

The principle of mediocrity, or the Copernican principle, suggests that Earth and its inhabitants are not unique. It posits that the conditions leading to life on Earth are not extraordinary but rather typical of planets in similar environments.

Extremophiles, microorganisms that thrive in extreme environments on Earth, demonstrate the adaptability of life. These organisms inhabit environments once considered inhospitable, raising the possibility of life in extreme extraterrestrial conditions.

The search for biosignatures, such as oxygen or methane, in the atmospheres of exoplanets within habitable zones is an essential aspect of the quest for extraterrestrial life. The presence of these markers could indicate the existence of life.

The Fermi Paradox highlights the apparent contradiction between the high probability of extraterrestrial civilizations in the galaxy and the lack of observed contact or communication with them. It prompts various hypotheses and theories about why we have not encountered extraterrestrial life.

The Drake Equation offers a framework for estimating the number of extraterrestrial civilizations in our galaxy. While its parameters are subject to debate, it provides a structured approach to assessing the potential prevalence of intelligent life.

The vast distances between stars and the limitations of current technology pose significant challenges to interstellar communication and travel. These challenges may explain our lack of direct contact with extraterrestrial civilizations.

SETI initiatives actively scan the cosmos for signals from intelligent civilizations. While no conclusive evidence has been found, the search continues, driven by the possibility of detecting signs of extraterrestrial intelligence.

Ongoing and future space exploration missions, including those to Mars, ocean worlds, and distant exoplanets, hold the potential to uncover evidence of past or present extraterrestrial life, providing valuable insights into this question.

The exploration of whether we are alone in the universe is a journey that combines scientific inquiry, technological advancements, and philosophical contemplation. It reflects humanity's enduring curiosity about our place in the cosmos and the potential for life to exist beyond the boundaries of our home planet.

3.4.2. Evolutionary Processes of Planets

The exploration of planetary evolution unveils the intricate and dynamic forces that shape celestial bodies over geological epochs. This understanding holds the key to unraveling the potential habitability of planets and the evolution of life on otherworldly landscapes.

Planetary genesis initiates within protoplanetary disks encircling nascent stars, where fine particles coalesce under gravitational pull, birthing planetesimals and, eventually, fully-fledged planets. A planet's composition, size, and orbital location at its genesis profoundly impact its subsequent evolution.

Differentiation is a pivotal phase wherein planets stratify into distinct layers. Heavy elements gravitate toward the core, while lighter materials accumulate in the outer regions, giving rise to cores, mantles, and crusts, each holding diverse geological narratives.

Volcanism emerges as a planetary narrator, spewing gases like water vapor and carbon dioxide into atmospheres, swaying climates, and sculpting landscapes with mountains, valleys, and volcanic vistas.

Tectonics, akin to Earth's plate tectonics, fuels the choreography of lithospheric plates, crafting continents, ocean basins, and recycling surface materials. These movements exert substantial influences on climates and the availability of essential elements.

Erosion and weathering etch the planetary canvas, driven by natural agents—wind, water, and ice—giving birth to landscapes as varied as river valleys, canyons, and sediment deposits, while redistributing vital minerals and nutrients.

Atmospheric metamorphosis undergoes transformations via geological and biological forces, with alterations in composition—such as oxygen accumulation via photosynthesis—holding sway over a planet's habitability.

Impacts and catastrophic episodes—wrought by cosmic visitors like comets and asteroids—leave their indelible marks, sculpting craters, triggering mass extinctions, and reshaping climates and geological profiles.

Climate oscillations are part of a planet's life story, driven by variables like orbital shifts, solar radiance, and shifts in atmospheric makeup. Understanding these ebbs and flows is pivotal in gauging habitability.

The presence and stability of a magnetic field play a protective role, guarding a planet's atmosphere and surface from the deleterious effects of solar and cosmic radiation. A planet's capacity to generate and sustain this shield is tied to its internal dynamics.

Biological influences hold a profound sway, as exemplified on Earth. Life's evolution and its intricate dance with the environment can both shape and be shaped by a planet's evolution.

The tapestry of planetary evolutionary processes, interwoven and complex, defines the conditions that chart a planet's potential for habitability and the promise of extraterrestrial life. Scrutinizing these narratives underpins our grasp of the cosmos' rich array of planets and their suitability as canvases for life beyond our terrestrial home.

3.4.3. Ideal Environments for Life in the Universe

The quest for extraterrestrial life leads scientists to explore a multitude of cosmic environments, each with its unique conditions that might support life as we know it.

Liquid water is a fundamental prerequisite for life, serving as a universal solvent. Environments featuring stable liquid water, such as oceans, rivers, and lakes, are primary targets in our search.

Ideal environments are characterized by temperate climates that maintain moderate temperatures, providing a comfortable range for life to thrive. These regions neither scorch with intense heat nor freeze with extreme cold.

Life relies on essential chemical elements and compounds like carbon, nitrogen, phosphorus, and sulfur. Environments rich in these nutrients, whether through organic or inorganic sources, offer fertile ground for life.

Reliable and stable energy sources are essential for life processes. Ideal environments provide steady energy from sources such as sunlight (for photosynthesis), geothermal activity, chemical reactions, or other energy-rich compounds.

Protection from harmful solar and cosmic radiation is critical for life's survival. Robust planetary atmospheres or magnetic shields shield against excessive radiation exposure.

Consistent climates, characterized by minimal temperature fluctuations and stable conditions, are favorable for life's adaptation and evolution.

Environments supporting diverse ecosystems with various niches and species increase the potential for complex life forms and ecological interactions.

Some level of geological activity, including tectonics and volcanism, replenishes essential minerals, gases, and nutrients, sustaining life over geological time scales.

Unique habitats such as hydrothermal vent systems on ocean floors, with geothermal heat and chemical-rich waters, nurture extremophiles and distinct ecosystems.

Oceans concealed beneath icy crusts, like those on Europa and Enceladus, hold the tantalizing possibility of liquid water and potential life.

Certain niche refuges, such as caves, underground tunnels, or subterranean aquifers, provide protection from extreme surface conditions.

Environments with significant gradients in electron availability (redox gradients) provide energy sources for various metabolic processes, supporting diverse life forms.

Environments boasting diverse chemical compositions and geological features offer opportunities for adaptation and evolution, fostering life's diversity.

Comprehending these ideal life-supporting environments expands our exploration horizons as we seek to understand the potential for life to thrive in the varied settings across the universe.

CHAPTER 4

Fermi Paradox and Communication Barriers

4.1. Fermi Paradox and Expectations of Intelligent Life

4.1.1. Origin of the Fermi Paradox

The Fermi Paradox, named after the renowned physicist Enrico Fermi, arises from the apparent contradiction between the high probability of extraterrestrial civilizations existing in the Milky Way galaxy and the lack of evidence or contact with such civilizations. Several hypotheses have been proposed to explain this perplexing paradox:

Rare Earth Hypothesis: This hypothesis suggests that Earth-like planets with conditions conducive to complex life forms are exceptionally rare in the universe. Even if life is common, the emergence of intelligent civilizations like ours may be an exceedingly rare occurrence.

Great Filter Theory: The Great Filter is a hypothetical stage or obstacle in the development of intelligent civilizations that prevents most from advancing to the point of interstellar communication or exploration. It could be a catastrophic event or a series of challenges that significantly reduce the number of civilizations.

Self-Destruction: Some theories posit that intelligent civilizations often self-destruct due to technological, environmental, or societal issues before they can develop advanced interstellar capabilities. This could explain the absence of extraterrestrial signals or contact.

Communication Challenges: Advanced civilizations may exist, but we might not detect their signals due to limitations in our technology or differences in communication methods. They may use communication channels or technologies beyond our current understanding.

The Zoo Hypothesis: This hypothesis proposes that advanced civilizations are aware of us but choose not to make contact, allowing humanity to evolve naturally without interference. They might be observing us discreetly, like a zookeeper observing animals in a wildlife sanctuary.

Interstellar Travel Constraints: Even if advanced civilizations exist, the vast distances between stars may make interstellar travel exceedingly difficult or impractical. This could limit the likelihood of direct contact.

Simulation Hypothesis: Some theories suggest that our reality is a computer simulation created by an advanced civilization. If this is the case, the lack of contact could be by design.

These are just a few of the many theories proposed to address the Fermi Paradox. Exploring these hypotheses and the broader implications of the paradox continues to be a fascinating topic of research and speculation in the field of astrobiology and astrophysics.

4.1.2. Prevalence of Intelligent Life

The Fermi Paradox raises questions not only about the existence of extraterrestrial civilizations but also about the prevalence of intelligent life in the universe. To address this aspect of the paradox, scientists have explored various factors that might influence the emergence of intelligent life.

The abundance of exoplanets, including those within the habitable zone of their parent stars, suggests that Earth-like planets might be common. This increases the potential for life to emerge elsewhere in the galaxy.

The existence of extremophiles, organisms capable of surviving in extreme environments on Earth, suggests that life may be more resilient and adaptable than previously thought. It raises the possibility of life in harsh cosmic conditions.

Advances in astrobiology have expanded our understanding of where life can exist. This research extends the potential habitable zones beyond what was initially considered.

The search for extraterrestrial intelligence (SETI) continues through radio telescopes and other methods. While no conclusive evidence has been found, ongoing efforts aim to detect potential signals from intelligent civilizations.

The Drake Equation, a probabilistic formula used to estimate the number of technologically advanced civilizations in our galaxy, takes into account factors such as star formation rates, planetary systems, and the probability of life evolving on suitable planets.

The anthropic principle suggests that the universe's physical constants and laws are fine-tuned to allow the emergence of life and intelligent observers.

Reevaluating the Rare Earth Hypothesis, which suggests that Earth-like planets are rare, is essential to understanding the prevalence of intelligent life.

The evolution of intelligent life may require specific conditions and a high level of complexity. Understanding the emergence of complex life forms is a critical aspect of this discussion.

The prevalence of intelligent life in the universe remains a topic of great interest and speculation, with ongoing scientific research and exploration aimed at shedding light on this intriguing aspect of the Fermi Paradox.

4.1.3. Expectations Versus Reality

The Fermi Paradox has led to a stark contrast between our expectations of encountering extraterrestrial civilizations and the reality of the situation. Several factors contribute to this disconnect:

Sci-Fi and Pop Culture: Popular culture, including science fiction literature and movies, often depicts frequent and elaborate encounters with intelligent extraterrestrial beings. These portrayals create high expectations that may not align with reality.

Vast Cosmic Distances: The immense distances between stars and galaxies make interstellar travel challenging. Even if advanced civilizations exist, the vastness of space might prevent direct contact.

Limitations of Technology: Our current technology and methods for searching for extraterrestrial signals have limitations. We may not yet possess the means to detect potential alien communications or civilizations.

Lack of Evidence: Despite extensive searches, we have not found conclusive evidence of extraterrestrial civilizations or signals. This contrasts with the expectation that we might have encountered them by now.

Nature of Intelligent Life: The evolution of intelligent life on Earth took billions of years. It's possible that other civilizations, if they exist, are at different stages of development, making contact less likely.

Communication Challenges: The methods and frequencies used by potential extraterrestrial civilizations may differ significantly from what we expect, making it difficult to detect their signals.

Possibility of Isolation: Advanced civilizations may choose to remain isolated for various reasons, such as avoiding interference with less developed species or protecting themselves from potential threats.

Complexity of Space: The complexity of the universe, with its countless stars, planets, and potential habitats, means that even if intelligent life is relatively common, it might be dispersed widely.

The gap between our expectations and the reality of the Fermi Paradox challenges our understanding of the cosmos and the potential existence of intelligent life beyond Earth.

4.2. Communication Barriers and Potential Causes

4.2.1. Galactic Messaging Challenges

When contemplating the search for extraterrestrial intelligence and the Fermi Paradox, it becomes apparent that several challenges complicate our ability to communicate with potential extraterrestrial civilizations within our galaxy:

Vast Distances: The Milky Way galaxy spans approximately 100,000 light-years, and stars and planets within it are separated by immense distances. Transmitting messages across such vast expanses presents logistical challenges.

Limitation of Light-Speed Communication: Messages sent using electromagnetic waves, which travel at the speed of light, require thousands of years to traverse the Milky Way. This time delay makes real-time communication with distant civilizations practically impossible.

Diversity of Civilizations: Assuming extraterrestrial civilizations exist, they may possess different levels of technological advancement and communication methods. Deciphering and interpreting messages from such diverse sources could be challenging.

Survivability of Messages: Messages sent into space must endure the harsh conditions of interstellar space, including cosmic radiation and micrometeoroid impacts. Ensuring the longevity of messages over cosmic timescales is a significant challenge.

Understanding Alien Languages: Even if we receive a signal from an extraterrestrial civilization, deciphering the language or communication method they use could be immensely challenging, given our lack of a shared linguistic or cultural context.

Message Complexity: Crafting a message that effectively conveys humanity's knowledge and intentions without misinterpretation is complex. Striking the right balance between simplicity and comprehensiveness is a crucial challenge.

Message Directionality: Determining where to send messages within the galaxy is a challenging decision. Do we target specific star systems, or should messages be sent in all directions in the hope that they might reach a receptive audience?

Response Timeframes: Even if we transmit messages and receive a response, the timeframes involved—potentially centuries or millennia—raise questions about continuity and maintaining a coherent dialogue.

Navigating these challenges in galactic messaging is essential for our ongoing efforts to establish contact with potential extraterrestrial civilizations and unravel the mysteries of the Fermi Paradox.

4.2.2. Earth's Communication Capacity

As we ponder the challenges of communicating with potential extraterrestrial civilizations, we must also consider Earth's communication capacity and its role in the search for intelligent life beyond our planet:

Technological Advancements: Earth has made significant technological advancements in the field of communication. The development of powerful radio telescopes and other equipment has enhanced our ability to send and receive signals from deep space.

Active SETI Initiatives: Some organizations and researchers have initiated Active SETI (Search for Extraterrestrial Intelligence) programs, deliberately sending messages into space in the hope of eliciting a response. These initiatives reflect our increasing communication capacity.

Passive SETI Observations: Beyond active messaging, Earth constantly listens for signals from potential extraterrestrial civilizations. Passive SETI observations use radio telescopes to monitor cosmic radio frequencies for any anomalous signals.

International Collaboration: The search for extraterrestrial intelligence is a global effort, with numerous countries and organizations collaborating to pool resources and expertise. This collaborative approach significantly enhances Earth's communication capacity.

Communication Protocols: Efforts are made to establish communication protocols and standards that would facilitate effective interstellar messaging. These protocols aim to ensure clarity and reduce the risk of misunderstanding.

Transmission Power: Earth possesses the capability to transmit signals with significant power, which increases the likelihood of these signals traveling vast distances through space.

Public Engagement: Public interest and engagement in the search for extraterrestrial intelligence contribute to Earth's communication capacity. Citizen science initiatives and outreach programs help monitor signals and analyze data.

Integration with Space Exploration: The coordination of SETI efforts with space exploration missions, such as those launched to study exoplanets, enhances our ability to discover potential extraterrestrial signals.

Earth's communication capacity plays a pivotal role in our pursuit of answers to the Fermi Paradox and our quest to establish contact with intelligent civilizations beyond our world. Advancements in technology, international cooperation, and the dedication of scientists and the public continue to expand our ability to communicate with the cosmos.

4.2.3. Thought Patterns of Unknown Beings

One of the profound challenges in our search for extraterrestrial intelligence and the resolution of the Fermi Paradox lies in deciphering the thought patterns and cognitive processes of entirely unknown beings:

Cultural and Evolutionary Differences: Extraterrestrial civilizations, if they exist, may have developed in radically different cultural and evolutionary contexts. Their thought patterns and values may diverge significantly from ours, making understanding their communication a complex task.

Non-Humanoid Intelligence: The possibility of non-humanoid, vastly different forms of intelligence complicates the task further. Sentient beings with alternative sensory perceptions or modes of cognition might express themselves in ways we can't readily grasp.

Communication Modalities: Our understanding of communication is rooted in human experiences and languages. Extraterrestrial civilizations may employ entirely different communication modalities—visual, auditory, chemical, or electromagnetic—that challenge our comprehension.

Conceptual Frameworks: Understanding an alien civilization's conceptual framework, including their mathematics, physics, and logic, is a daunting endeavor. Concepts we consider fundamental might not be universal.

Context and Symbolism: Extraterrestrial messages might contain symbols or references that are utterly unfamiliar to us. Deciphering these symbols and understanding the cultural context behind them could be a formidable task.

Shared Knowledge: Effective communication often relies on shared knowledge and common reference points. Extraterrestrial civilizations might lack these shared foundations, requiring us to build understanding from the ground up.

Misinterpretation Risks: The risk of misinterpreting messages or intentions is substantial. Even well-intentioned communication attempts could lead to misunderstandings due to differing cognitive frameworks.

Psychological Impact: The mere discovery of extraterrestrial intelligence and attempts to communicate with unknown beings may have profound psychological and cultural impacts on humanity, affecting our thought patterns and worldviews.

Interdisciplinary Collaboration: Successfully deciphering the thought patterns of unknown beings requires interdisciplinary collaboration across fields like linguistics, psychology, anthropology, and astrophysics.

Navigating these complexities in understanding the thought patterns of unknown beings is a critical aspect of our quest to communicate with potential extraterrestrial civilizations and unravel the mysteries of the Fermi Paradox.

4.3. Developmental Processes of Extraterrestrial Civilizations

4.3.1. Humanity's History and Evolution

To explore the Fermi Paradox and our search for extraterrestrial intelligence, it's essential to consider the history and evolution of humanity:

Emergence of Homo Sapiens: Homo sapiens, modern humans, emerged in Africa approximately 200,000 years ago. Understanding our own evolution and history helps us contemplate the potential evolution of intelligent life elsewhere.

Cultural and Technological Evolution: Over millennia, humans have advanced culturally and technologically, developing complex societies, languages, and technologies. This history informs our expectations of the developmental stages of potential extraterrestrial civilizations.

The Scientific Revolution: The Scientific Revolution, beginning in the 16th century, marked a pivotal point in human history. It accelerated our understanding of the natural world and laid the groundwork for advanced technology.

Technological Acceleration: The past century has seen exponential growth in technology, with breakthroughs in space exploration, computing, and communication. This rapid advancement might serve as a reference point for assessing the potential technological trajectories of other civilizations.

Space Age and Space Exploration: The Space Age, which began with the launch of Sputnik in 1957, opened new possibilities for space exploration. Our exploration of the cosmos provides insights into the challenges and opportunities of interstellar travel.

Communication Revolution: The development of the internet and global communication networks has transformed how we exchange information. This revolution may have implications for our own communication with potential extraterrestrial civilizations.

Astrobiology and SETI: The emergence of astrobiology and the Search for Extraterrestrial Intelligence (SETI) as scientific disciplines reflect our curiosity and dedication to understanding our place in the universe.

Cultural and Ethical Considerations: As we explore the cosmos and contemplate contact with potential extraterrestrial beings, we grapple with ethical and cultural questions, such as our responsibility as stewards of Earth and the potential impact of contact on our societies.

Technological Evolution and Future Prospects: Considering our rapid technological evolution and the potential for future breakthroughs, we contemplate the development and capabilities of advanced extraterrestrial civilizations.

Reflecting on humanity's history and evolution provides valuable context for our search for intelligent life beyond Earth and the Fermi Paradox. It helps us imagine the trajectories and potential challenges faced by other civilizations in the cosmos.

4.3.2. How Another Civilization Might Evolve

When contemplating the Fermi Paradox and the potential existence of extraterrestrial civilizations, it's instructive to consider how another civilization might evolve:

Emergence of Life: Like Earth, life on another planet may have originated in a primordial soup, evolving from simple single-celled organisms to complex, multicellular life forms.

Cultural and Technological Progress: Over time, a civilization could develop its culture and technology, potentially leading to the creation of advanced societies, languages, and sciences.

Technological Advancements: As with humanity, technological breakthroughs might propel the civilization forward. The development of tools, agriculture, and industrialization could be significant milestones.

Space Exploration: A key turning point might come with the civilization's exploration of space. Advancements in propulsion, robotics, and artificial intelligence could enable interplanetary and interstellar exploration.

Contact with Extraterrestrial Life: If the civilization encounters other life forms, the experience could profoundly shape its development, fostering cooperation or competition.

Resource Management: Managing finite resources and environmental challenges could influence the civilization's survival and development. Sustainability and responsible resource use might become critical.

Cultural Evolution: Cultural norms and values could evolve, impacting the civilization's decisions regarding ethics, governance, and societal structures.

Scientific Advancements: The civilization's understanding of physics, biology, and other scientific disciplines might advance, enabling new technologies and innovations.

Interstellar Travel: If feasible, interstellar travel could open up opportunities for colonization and interaction with other star systems, potentially leading to the spread of the civilization throughout the galaxy.

Communication Technologies: The development of advanced communication technologies, such as quantum communication or other interstellar messaging methods, could facilitate contact with other civilizations.

Ethical Considerations: Ethical dilemmas and philosophical questions may arise as the civilization faces choices about its interactions with other intelligent life forms.

Survival and Evolution: The civilization's ability to navigate challenges, adapt, and survive could be a crucial factor in its long-term evolution.

Considering how another civilization might evolve offers valuable insights into our search for extraterrestrial intelligence and the Fermi Paradox. It encourages us to imagine the diverse trajectories that intelligent life forms might follow throughout the cosmos.

4.3.3. The Future of Technological Civilizations

As we ponder the Fermi Paradox and the potential existence of extraterrestrial civilizations, it's essential to consider the future trajectories of advanced technological civilizations:

Sustainability and Resource Management: Advanced civilizations must grapple with the sustainable management of resources to ensure their long-term survival. Strategies for responsible resource use and environmental stewardship become paramount.

Technological Advancements: The continued development of technology plays a central role in a civilization's future. Scientific breakthroughs, engineering innovations, and technological evolution drive progress.

Interstellar Expansion: The concept of expanding beyond a civilization's home planet or star system may become a reality. Colonization of other celestial bodies, star systems, and even galaxies could be on the horizon.

Communication with Other Civilizations: Advanced civilizations may invest in communication technologies designed for interstellar messaging.

Initiatives like METI (Messaging to Extraterrestrial Intelligence) could play a significant role.

Interactions with Alien Life: The possibility of encountering extraterrestrial life, whether microbial or intelligent, could reshape a civilization's future. Ethical considerations and diplomatic strategies for interaction are vital.

Technological Singularities: Some civilizations may approach technological singularities, wherein artificial intelligence and advanced technology accelerate beyond human control or comprehension.

Energy Sources and Sustainability: Developing sustainable and efficient energy sources becomes increasingly important for the long-term viability of advanced civilizations.

Cultural Evolution: The cultural and societal evolution of civilizations can be dynamic. Changes in values, ethics, governance, and social structures may shape their futures.

Exploration of Cosmic Phenomena: Advanced civilizations may dedicate resources to exploring cosmic phenomena, such as black holes, dark matter, and other mysteries of the universe.

Existential Risks and Mitigation: Identifying and mitigating existential risks—threats that could imperil the civilization's survival—becomes a priority. This includes concerns about asteroid impacts, supervolcanoes, and potential cosmic disasters.

Integration of Biological and Synthetic Life: Advancements in biotechnology and synthetic biology might lead to the integration of biological and synthetic life forms, blurring the boundaries between biology and technology.

Ethical Frameworks and Governance: Developing ethical frameworks and governance structures for the responsible use of technology and the management of complex societal issues becomes essential.

The future of technological civilizations is a complex and multifaceted topic, offering numerous possibilities and challenges. Considering these potential trajectories enriches our understanding of the Fermi Paradox and the diverse paths intelligent life may follow throughout the universe.

4.4. Implications Raised by the Fermi Paradox

4.4.1. Space-Time Tunnel and Communication

Exploring potential solutions to the Fermi Paradox and the challenges of interstellar communication, one intriguing concept that has been considered is the idea of a space-time tunnel:

Wormholes and Space-Time Tunnels: Wormholes are theoretical passages through space-time that could potentially connect distant regions of the universe. While their existence remains speculative, they are a topic of scientific inquiry.

Instantaneous Communication: If space-time tunnels like wormholes were to exist and could be harnessed, they might enable nearly instantaneous communication across vast cosmic distances.

Theoretical Feasibility: The feasibility of space-time tunnels hinges on complex theories in physics, including general relativity and quantum mechanics. Researchers explore the mathematical and theoretical underpinnings of these concepts.

Stability and Traversability: One of the challenges associated with wormholes is their stability and traversability. Creating a stable tunnel without destructive gravitational forces is a significant theoretical hurdle.

Advanced Technology Requirements: Harnessing space-time tunnels would likely require highly advanced technology beyond our current capabilities, making it a topic of speculation rather than practical application at present.

Extraterrestrial Potential: If space-time tunnels exist and advanced extraterrestrial civilizations have discovered how to navigate them, they could facilitate communication and travel among star systems.

Interstellar Exploration: Beyond communication, the existence of space-time tunnels might also open up possibilities for interstellar exploration and colonization.

Theoretical Limits: It's important to note that the concept of space-time tunnels remains highly speculative and theoretical, with no empirical evidence to support their existence as of now.

The notion of space-time tunnels presents an intriguing avenue for considering how advanced civilizations might overcome the vast cosmic distances and communicate across the galaxy. While it remains speculative, the exploration of such concepts broadens our perspective on the Fermi Paradox.

4.4.2. Unpredictability of the Future

When contemplating the Fermi Paradox and the challenges of understanding extraterrestrial intelligence, one significant factor to consider is the inherent unpredictability of the future:

Complex Systems: The evolution of intelligent civilizations and their potential for space exploration are influenced by a multitude of complex and interconnected factors. These include cultural, environmental, technological, and sociopolitical dynamics.

Unforeseen Events: The course of history on Earth has been shaped by unforeseen events and contingencies. Similarly, extraterrestrial civilizations might face unpredictable challenges or opportunities that affect their development.

Technological Surprises: Breakthroughs in science and technology often occur unexpectedly. The emergence of entirely new fields of research or unexpected inventions could significantly alter the trajectory of a civilization.

Cultural Shifts: The evolution of cultures and societies can be unpredictable. Shifts in values, beliefs, and worldviews can lead to profound changes in a civilization's goals and priorities.

Exogenous Threats: Civilizations might encounter exogenous threats, such as cosmic disasters or the discovery of powerful and potentially dangerous cosmic phenomena, which could reshape their plans and survival strategies.

Interactions with Other Civilizations: Encounters with other civilizations, if they occur, could be highly unpredictable. The nature of these interactions— cooperation, competition, or isolation—may depend on unique circumstances.

Technological Singularities: The development of advanced artificial intelligence or other technologies could lead to technological singularities, beyond which the future becomes difficult to predict.

Cultural and Ethical Evolution: As cultures and societies evolve, their ethical frameworks and values may undergo significant changes, impacting their approach to exploration and contact with other intelligent beings.

Random Events: Chance events and random factors can play a role in a civilization's trajectory. These events are, by their nature, unpredictable.

Adaptation to Challenges: A civilization's ability to adapt and respond to unforeseen challenges may determine its long-term survival and influence its future development.

Acknowledging the unpredictability of the future is a humbling aspect of contemplating the Fermi Paradox. While we can develop scenarios and hypotheses, the course of intelligent civilizations, including potential extraterrestrial ones, is ultimately uncertain and influenced by a myriad of factors, many of which are beyond our current knowledge and foresight.

4.4.3. Humanity's Path and Vision Towards the Stars

As we delve into the Fermi Paradox and the quest to understand extraterrestrial intelligence, it's crucial to consider humanity's own path and vision for venturing into the cosmos:

Historical Space Exploration: Humanity's journey into space began with the launch of artificial satellites, such as Sputnik in 1957, and continued with historic events like the Apollo Moon landings. These milestones laid the foundation for future exploration.

Space Agencies and Collaborations: Nations and space agencies, including NASA, ESA, Roscosmos, and others, have conducted extensive space missions, explored the planets of our solar system, and sent spacecraft beyond the heliosphere.

International Space Station (ISS): The ISS represents a remarkable international collaboration for the peaceful exploration of space. It serves as a platform for scientific research and a symbol of humanity's commitment to space.

Mars Exploration: The exploration of Mars has become a focal point, with robotic missions like the Mars rovers paving the way for future human missions to the Red Planet.

Private Space Industry: The emergence of private space companies, such as SpaceX, Blue Origin, and others, has revolutionized space access, potentially enabling more frequent and affordable space exploration.

Vision for Moon and Mars: Ambitious plans for lunar and Martian exploration, including Artemis and the aspiration to establish a sustainable presence on the Moon and send humans to Mars, reflect humanity's vision for interplanetary expansion.

Search for Life: Space missions, such as the study of Mars for signs of past or present life and the exploration of icy moons like Europa and Enceladus for subsurface oceans, are driven by the quest to understand the potential for life beyond Earth.

Interstellar Ambitions: Concepts like the Breakthrough Starshot initiative aim to send small probes to neighboring star systems within our lifetime, demonstrating humanity's aspirations for interstellar exploration.

Technological Advancements: Advances in propulsion, materials science, artificial intelligence, and other fields continue to shape the future of space exploration.

Cosmic Vision: Humanity's cosmic vision encompasses the search for extraterrestrial intelligence, the study of exoplanets, and the exploration of cosmic phenomena.

Challenges and Ethical Considerations: The pursuit of space exploration raises ethical questions about planetary protection, resource utilization, and our responsibilities as stewards of the cosmos.

Public Engagement: Public interest and engagement in space exploration and science communication play a crucial role in shaping humanity's path and vision for the stars.

Considering humanity's journey into space and its vision for the future provides valuable context for our quest to understand potential extraterrestrial civilizations and the mysteries of the Fermi Paradox. It reflects our collective aspiration to explore and discover the cosmos beyond our home planet.

CHAPTER 5

Communication Initiatives and SETI

5.1. SETI: The Quest for Extraterrestrial Intelligent Life

5.1.1. Foundations of SETI

The Search for Extraterrestrial Intelligence, or SETI, is a scientific pursuit with profound implications. Its foundations rest upon the idea of detecting signs of intelligent life beyond our planet. This endeavor has evolved over the years and is marked by several critical components.

One of the cornerstones of SETI is radio astronomy. This discipline forms the basis for listening to potential artificial signals from civilizations elsewhere in the cosmos. It gained prominence in the mid-20th century, shaping the course of SETI research.

A pivotal moment in the history of SETI was the formulation of the Drake Equation by Dr. Frank Drake in

1961. This equation is a framework that attempts to estimate the number of communicative extraterrestrial civilizations within our galaxy. It considers factors like star formation rates, planetary systems, and the emergence of life on habitable planets.

Project Ozma, led by astronomer Frank Drake in 1960, is regarded as the first systematic attempt to detect extraterrestrial radio signals. Focusing on stars like Tau Ceti and Epsilon Eridani, it laid the groundwork for subsequent SETI initiatives, demonstrating the feasibility of such searches.

The Green Bank Conference in 1961 marked a turning point in the formalization of SETI as a scientific discipline. It brought together scientists to discuss the pursuit of extraterrestrial intelligence, setting the stage for international collaboration and research efforts.

Advancements in technology have played a crucial role in the evolution of SETI. The development of radio interferometry and the utilization of observatories like Arecibo and the Very Large Array have expanded the scope and precision of SETI endeavors.

In recent years, the field has broadened its horizons to explore alternative modes of communication beyond radio signals, including optical signals, laser transmissions, and broadband emissions. The quest for extraterrestrial intelligence continues to captivate both scientists and the public, engaging millions of volunteers in projects like SETI@home.

International cooperation remains a cornerstone of SETI, with observatories and organizations worldwide working together to scan the skies for signals. This collaboration ensures comprehensive coverage of the cosmos and enhances the chances of a potential discovery.

The search for anomalies in signals that might indicate artificial origins is a fundamental aspect of SETI research. Scientists scrutinize patterns, frequencies, and other characteristics, seeking signals that deviate from natural phenomena.

Astrobiology and the discovery of exoplanets within habitable zones have rekindled enthusiasm for SETI, emphasizing the potential for life elsewhere in the universe. Theoretical frameworks guide SETI researchers in contemplating the motivations and methods of communication employed by extraterrestrial civilizations.

Ethical and cultural considerations are also integral to SETI investigations. Questions surrounding the responsible transmission of messages from Earth and the potential consequences of contact underscore the importance of these aspects in the search for extraterrestrial intelligence.

In conclusion, the foundations of SETI are rooted in scientific rigor and a relentless pursuit of answers to one of humanity's most profound questions: Secrets of the Cosmos As technology advances and international collaboration deepens, the quest for signs of intelligent life beyond Earth continues to be a source of fascination and inspiration for us all.

5.1.2. The Scope of Communication Initiatives

Within the realm of SETI, the scope of communication initiatives is vast and multifaceted. These efforts extend beyond radio signals and encompass a variety of strategies and considerations in the search for extraterrestrial intelligence.

At its core, the search for extraterrestrial intelligence revolves around deciphering potential signals or evidence of intelligent life. While radio astronomy has historically been the primary avenue for this search, the field has expanded to explore various other modes of communication.

Optical signals represent one alternative avenue in the quest for extraterrestrial intelligence. Scientists look for patterns of light that may indicate intentional signaling. Laser transmissions, in particular, offer a promising means of communication across interstellar distances.

Broadband emissions are another facet of communication initiatives. By scanning a broader range of frequencies, scientists aim to detect any anomalies or patterns that may be indicative of intelligent origins. The search criteria are adaptable, allowing for the exploration of diverse signal types.

Public engagement plays a significant role in communication initiatives. Projects like SETI@home have harnessed the computational power of volunteers worldwide to process vast amounts of data. This crowd-sourced approach enhances the efficiency of signal analysis and broadens the community engaged in the search.

International collaboration remains a hallmark of communication efforts in SETI. Coordinated observations across multiple observatories and research institutions ensure a comprehensive scan of the cosmos. The sharing of data and resources bolsters the collective potential for discovery.

In parallel with signal detection, researchers grapple with the profound ethical and cultural dimensions of communication with potential extraterrestrial civilizations. The responsible transmission of messages from Earth and the possible consequences of contact remain subjects of ongoing debate and consideration.

As technology continues to advance, so too does the scope of communication initiatives within SETI. Scientists and researchers explore new methods, refine search criteria, and collaborate on an unprecedented scale. The quest to decipher the cosmic conversation and unravel the mysteries of potential extraterrestrial intelligence remains a journey of profound scientific and philosophical significance.

5.1.3. Signals from the Cosmos

In the relentless pursuit of extraterrestrial intelligence, scientists meticulously examine signals emanating from the cosmos. These signals, often shrouded in cosmic noise, represent potential indicators of intelligent life beyond Earth. SETI's approach to decoding these signals is marked by scientific rigor and adaptability.

Radio Signals: Radio astronomy has been at the forefront of SETI efforts, focusing on specific radio frequencies where artificial signals might stand out. Researchers analyze radio emissions for patterns, anomalies, or repetitions that could signify intentional transmission. This avenue has yielded many compelling candidates for further investigation.

Optical Signals: Optical signals, characterized by patterns of light, have gained prominence in recent years. Laser emissions, in particular, are scrutinized for signs of intelligent modulation. The search extends to the visual part of the electromagnetic spectrum, where scientists explore the possibility of optical communication.

Broadband Emissions: Broadband signals encompass a wide range of frequencies, providing a comprehensive view of the electromagnetic spectrum. SETI investigations consider the potential for intelligent signals across this spectrum, including both narrowband and broadband emissions. The adaptability of search criteria allows for the exploration of diverse signal types.

Artificial versus Natural: One of the fundamental challenges in decoding signals from the cosmos is distinguishing between artificial and natural sources. Researchers meticulously examine the characteristics of signals, aiming to discern intentional patterns that are unlikely to arise from natural cosmic phenomena.

Public Involvement: Public engagement remains a pivotal aspect of SETI's signal analysis endeavors. Projects like SETI@home leverage the computational power of volunteers worldwide, enabling the processing of vast datasets. This collective effort enhances the efficiency of signal analysis and extends the reach of the search.

Global Collaboration: The search for signals from the cosmos is a global endeavor. International collaboration among observatories, research institutions, and experts fosters a unified approach to data collection and analysis. This collaborative spirit enhances the potential for detecting extraterrestrial intelligence.

Ethical and Philosophical Considerations: Beyond the scientific realm, SETI confronts profound ethical and philosophical questions. Researchers grapple with the responsibility of transmitting messages from Earth and the potential consequences of contact. These considerations shape the ethical framework within which SETI operates.

The journey to decipher signals from the cosmos is an ongoing one, characterized by both excitement and patience. As technology advances and international cooperation deepens, scientists and researchers continue to refine their methods and broaden their search criteria. The quest to unveil the cosmic conversation and uncover signs of intelligent life in the universe remains a testament to human curiosity and the enduring pursuit of knowledge.

5.2. Earth's Communication Efforts into Space

5.2.1. Voyager Golden Record: Humanity's Message

The Voyager Golden Record stands as a testament to humanity's boundless curiosity and its aspiration to communicate with potential extraterrestrial civilizations. Launched aboard the Voyager 1 and Voyager 2 spacecraft in 1977, this unique artifact carries a message from Earth to the far reaches of the cosmos.

Crafting the Record: A team led by Dr. Carl Sagan was tasked with curating the content of the Voyager Golden Record. Their aim was to encapsulate the essence of humanity and Earth's diversity within a 12-inch, gold-plated copper disk. The record was designed to endure for billions of years, providing an enduring snapshot of our civilization.

Multimedia Compendium: The Voyager Golden Record contains a treasure trove of audio and visual content. It includes greetings in 55 languages, music from various cultures, and a selection of natural sounds from Earth. The audio portion is accompanied by a visual guide etched onto the record's surface, providing instructions on how to play it.

Sounds of Earth: The audio selection is a symphony of Earth's rich sonic tapestry. It features classical compositions, traditional music, and iconic songs like "Johnny B. Goode" by Chuck Berry. Natural sounds, such as thunder, wind, and animal calls, were also included to provide a sense of the planet's environment.

The "Hello" Message: A special greeting from the children of Planet Earth opens the Voyager Golden Record. This message is followed by welcoming words in numerous languages, reflecting the global diversity of humanity.

Pictorial Messages: The visual component of the record comprises 115 images, offering a glimpse into Earth's landscapes, species, and cultures. These images are encoded in analog form, with each frame carefully selected to convey aspects of our world.

Cosmic Ambassadors: The Voyager 1 and Voyager 2 spacecraft, each carrying a Golden Record, have now ventured beyond our solar system. They serve as humanity's ambassadors, hurtling through interstellar space. While the likelihood of them encountering extraterrestrial life remains exceedingly low, the Voyager Golden Record stands as an enduring testament to our curiosity and desire to connect with the cosmos.

The Voyager Golden Record remains a symbol of humanity's quest for knowledge and its aspiration to reach out to the unknown. It is a reminder that our existence is part of a vast universe, and the record's journey through the cosmos serves as a beacon of our curiosity and the universality of human culture.

5.2.2. Radio Waves to Space

In the quest to communicate with potential extraterrestrial civilizations, radio waves have stood as a steadfast bridge between Earth and the cosmos. This method of transmission has played a pivotal role in our efforts to reach out to the unknown realms of the universe.

Radio Signals as Messengers: Radio waves, with their ability to traverse vast cosmic distances, have been at the heart of our extraterrestrial communication endeavors. Scientists have meticulously scrutinized the radio frequency spectrum, searching for patterns or emissions that may carry the hallmark of intelligent origin.

The Significance of Frequencies: The selection of specific radio frequencies is a critical aspect of this communication strategy. Certain frequencies are chosen for their potential to propagate through the cosmos with minimal interference. The focus on these frequencies enhances the chances of a detectable signal reaching its intended cosmic audience.

Narrowband versus Broadband: SETI research encompasses both narrowband and broadband radio signals. Narrowband signals are characterized by a focused, narrow range of frequencies, while broadband signals cover a broader spectrum. This flexibility allows scientists to explore various modes of potential communication.

Deciphering the Patterns: One of the key challenges in the analysis of radio signals is deciphering the patterns within the noise of the cosmos. Researchers meticulously examine these signals, looking for modulations, repetitions, or other characteristics that may signify intentional transmission.

Public Engagement and Crowdsourced Computing: Public participation is a vital aspect of radio wave-based communication initiatives. Projects like SETI@home harness the computational power of volunteers worldwide to process immense datasets. This collaborative approach accelerates signal analysis and broadens the search's reach.

Global Collaboration: The pursuit of radio wave-based communication is a global endeavor, marked by international collaboration. Observatories, research institutions, and experts worldwide pool their resources and expertise to survey the skies for potential signals.

Ethical and Philosophical Contemplations: Beyond the technical aspects, the use of radio waves to communicate with the cosmos raises profound ethical and philosophical questions. Researchers grapple with the implications of transmitting messages from Earth and contemplate the potential consequences of contact. These considerations guide the ethical framework within which SETI operates.

In our cosmic outreach efforts, radio waves remain a beacon of hope, a medium through which we extend our hand to the unknown. With each moment of collaboration and each advancement in technology, humanity's quest to communicate across the vast cosmic expanse continues to evolve, driven by our boundless curiosity and the enduring desire to connect with potential cosmic neighbors.

5.2.3. Communicating into the Unknown

As we cast our gaze into the cosmic abyss in search of potential extraterrestrial civilizations, the art of communicating into the unknown takes center stage. This endeavor is marked by both the methods and the profound considerations that underpin our quest to reach out to distant and enigmatic worlds.

Radiowaves: Radiowaves have historically been the primary medium for transmitting intentional signals across the cosmos. SETI researchers meticulously analyze radio frequencies for patterns or anomalies that might signify intelligent communication. The adaptability of radiowaves allows for the exploration of diverse signals.

Optical Signals: The exploration of optical signals, particularly laser transmissions, has gained prominence in recent years. Scientists scrutinize the spectrum of visible light for encoded messages that may stand out against the backdrop of the universe's natural emissions.

Broadband Emissions: The spectrum of electromagnetic frequencies offers an expansive canvas for communication. SETI investigations encompass both narrowband and broadband emissions, aiming to detect any signals that deviate from typical cosmic phenomena.

Deciphering Intent: One of the critical challenges in communicating into the unknown is deciphering the intent behind any detected signals. Researchers examine the characteristics of signals, looking for patterns or modulations that suggest intelligent origins.

The Power of Public Involvement: The search for extraterrestrial intelligence is not confined to the realm of scientists alone. Public engagement plays a significant role, with initiatives like SETI@home enlisting the computational power of volunteers worldwide to process immense datasets. This collective effort enhances the efficiency of signal analysis.

Global Collaboration: The pursuit of communication into the unknown is a global endeavor, marked by international cooperation. Observatories, research institutions, and experts worldwide coordinate their efforts to scan the skies for signals. The sharing of data and resources strengthens the collective potential for discovery.

Ethical and Philosophical Dimensions: Beyond the scientific realm, communication into the unknown raises profound ethical and philosophical questions. Researchers grapple with the responsibility of transmitting messages from Earth and contemplate the potential consequences of contact. These considerations are integral to the ethical framework within which SETI operates.

As we venture further into the cosmos, the art of communicating into the unknown remains an intricate tapestry of science, technology, and philosophy. With each technological leap and each moment of international collaboration, humanity inches closer to unraveling the cosmic conversation and potentially encountering signs of intelligent life beyond our home planet.

5.3. Radio Telescopes and SETI Projects

5.3.1. Arecibo and the Ears of Earth

The Arecibo Observatory in Puerto Rico once stood as an iconic symbol of humanity's ears to the cosmos. Nestled in the lush hills, this colossal radio telescope played a pivotal role in our quest to listen for signals from extraterrestrial civilizations.

A Legendary Observatory: Arecibo was not just an observatory; it was a scientific legend. Its colossal dish, stretching over 300 meters in diameter, was the world's largest single-dish radio telescope for decades. It boasted a unique spherical design, which allowed it to point in any direction, capturing radio waves from the deepest recesses of space.

Listening to the Stars: Arecibo's primary mission was to collect and analyze radio signals from the universe. It scanned the cosmos for signs of intelligent life by examining narrowband and broadband radio emissions. The observatory's sensitivity made it a powerful instrument for detecting even faint and distant signals.

The Search for Extraterrestrial Intelligence: Arecibo played a significant role in the Search for Extraterrestrial Intelligence (SETI). It contributed to numerous projects aimed at scanning the skies for artificial radio signals. Researchers at the observatory combed through vast amounts of data in their quest to find evidence of cosmic neighbors.

A Source of Scientific Insight: Beyond its SETI efforts, Arecibo was a versatile research facility. It was instrumental in studying pulsars, asteroids, planets, and the Earth's atmosphere. Its radar capabilities enabled detailed planetary exploration, including radar maps of Venus and Mercury.

The End of an Era: Sadly, the Arecibo Observatory met its demise in 2020 when the suspended platform, housing crucial equipment, collapsed. This heartbreaking event marked the end of an era for one of the world's most iconic observatories.

A Legacy Lives On: While the physical structure of Arecibo may be gone, its legacy endures in the hearts of scientists and space enthusiasts. The data collected over decades of operation continues to fuel scientific discoveries and inspire the next generation of astronomers.

Listening to the cosmos through the "ears" of Arecibo was a profound scientific endeavor. Although the observatory is no longer with us, its contributions to our understanding of the universe and its ongoing influence on the field of radio astronomy ensure that Arecibo's legacy will echo through the annals of scientific history.

5.3.2. The Green Bank Telescope and Other Endeavors

In the pursuit of unraveling the cosmic mysteries and listening for signals from the depths of space, the Green Bank Telescope (GBT) in West Virginia emerges as a prominent figure among the various endeavors to explore the universe's radio waves.

The Green Bank Telescope: The GBT, perched within the tranquil surroundings of the National Radio Quiet Zone, is a testament to human ingenuity in the realm of radio astronomy. Standing at 485 feet tall, it boasts a dish diameter of 100 meters, making it one of the largest fully steerable radio telescopes on Earth.

A Radio-Silent Sanctuary: The GBT's location within the National Radio Quiet Zone ensures minimal radio interference, allowing for the detection of even faint and distant cosmic signals. This pristine radio-silent sanctuary is crucial for conducting sensitive observations.

Cosmic Eavesdropping: The primary mission of the GBT is to eavesdrop on the radio emissions emanating from celestial objects. Researchers employ its immense sensitivity to explore a myriad of cosmic phenomena, from the study of distant galaxies to the search for pulsars and other exotic celestial objects.

A Dynamic Observatory: The GBT's versatility extends to its ability to observe a wide range of radio frequencies. This adaptability enables scientists to explore different aspects of the universe, from analyzing the cool hydrogen gas between galaxies to investigating the chemistry of interstellar space.

Beyond the GBT: While the GBT stands out as a flagship observatory, it is part of a broader network of radio telescopes and scientific endeavors. Collaborative efforts extend across the globe, with observatories like the Very Large Array (VLA) in New Mexico and international projects enhancing our collective capacity to explore the cosmos.

The Quest for the Unknown: The Green Bank Telescope, along with its counterparts and collaborators, embodies humanity's enduring quest to understand the universe. The information gleaned from these scientific endeavors broadens our horizons, fosters new discoveries, and kindles the eternal flame of curiosity that propels us deeper into the cosmic unknown.

As radio telescopes like the Green Bank Telescope continue to scan the skies, they stand as beacons of scientific exploration, offering a glimpse into the enigmatic realms of the universe. The pursuit of knowledge, underpinned by the ceaseless quest to decipher cosmic radio signals, remains an enduring testament to the indomitable spirit of human curiosity.

5.3.3. The Significance of Space Observations

Space-based observations hold profound significance within the realm of astronomy and astrophysics. These observations, conducted beyond Earth's atmosphere, have profoundly impacted our comprehension of the universe and remain pivotal in the pursuit of cosmic exploration.

Escaping Earth's Atmosphere: Space-based observations offer an unparalleled advantage by liberating scientists from Earth's atmosphere. While essential for life, our atmosphere can distort and filter incoming cosmic radiation, impairing the precision of astronomical observations. Space telescopes circumvent this limitation, providing pristine views of the cosmos.

Hubble Space Telescope: The Hubble Space Telescope, an iconic example, has delivered breathtaking imagery and invaluable scientific data since its 1990 launch. Its observations have deepened our comprehension of distant galaxies, nebulae, and the universe's age. Among Hubble's breakthroughs is the determination of the Hubble Constant, a pivotal cosmological parameter.

Beyond the Visible Spectrum: Space-based observatories extend our vision beyond the visible light spectrum. Equipped with specialized instruments, these telescopes capture data in ultraviolet, X-ray, and infrared wavelengths, unveiling cosmic phenomena concealed from our naked eyes.

The Spitzer Space Telescope: Spitzer, an infrared observatory, explored the universe's infrared realm, revealing insights into star, planet, and galaxy formation, as well as the detection of exoplanets and interstellar dust composition.

Chandra X-ray Observatory: Chandra, an X-ray observatory, has probed the high-energy universe, unraveling the dynamics of exploding stars, matter behavior around black holes, and X-ray emissions from galaxy clusters, illuminating cosmic forces.

Future Space Observatories: The significance of space-based observations continues to ascend with the launch of new observatories. The James Webb Space Telescope (JWST), poised to succeed Hubble, promises to revolutionize our comprehension of the early universe, exoplanets, and more.

International Collaboration: Numerous space observatories result from international cooperation, with multiple space agencies pooling resources and expertise. These partnerships amplify scientific potential and foster global collaboration in the pursuit of knowledge.

The Unending Odyssey: Space-based observations epitomize an unceasing voyage of discovery. As humanity ventures deeper into the cosmos, these observatories remain guiding lights of exploration, granting glimpses into the universe's farthest reaches and delivering answers to enduring questions about our cosmic origins. They represent the indomitable human spirit of exploration and an insatiable thirst for knowledge.

5.4. The Historical and Future Role of SETI

5.4.1. Early SETI Initiatives

The genesis of the Search for Extraterrestrial Intelligence (SETI) traces back to early initiatives marked by visionary scientists and the pioneering quest to intercept possible signals from intelligent beings beyond Earth.

Radio Pioneers: Early SETI initiatives drew inspiration from the work of visionaries like Guglielmo Marconi and Nikola Tesla, who pioneered radio communication. Scientists realized that radio waves, with their ability to traverse vast distances, could serve as a universal medium for potential interstellar communication.

Project Ozma: The foundational moment for SETI arrived with Project Ozma in 1960, led by astronomer Frank Drake. This groundbreaking project used the National Radio Astronomy Observatory's radio telescope in Green Bank, West Virginia, to listen for signals from Tau Ceti and Epsilon Eridani, two nearby star systems. While no signals were detected, Project Ozma marked the birth of systematic SETI efforts.

The Drake Equation: Frank Drake introduced the Drake Equation in 1961 as a framework for estimating the number of civilizations in our galaxy capable of communicating with us. This equation synthesized various factors, including the rate of star formation and the likelihood of life emerging on habitable planets.

The Golden Age of Radio SETI: The 1960s and 1970s witnessed the emergence of various radio-based SETI programs. Initiatives like the Cyclops Study explored the feasibility of large-scale radio telescopes dedicated to SETI. Although funding challenges and technical limitations persisted, the pursuit of cosmic signals continued.

The Wow! Signal: In 1977, the Ohio State University's Big Ear radio telescope detected a strong radio signal from the direction of the Sagittarius constellation. This signal, famously dubbed the "Wow! Signal," remains one of the most enigmatic episodes in SETI history, as its source was never conclusively identified.

Technology Advancements: Early SETI initiatives laid the groundwork for future efforts, fostering technological advancements in radio astronomy, signal processing, and data analysis. These developments paved the way for more ambitious and sophisticated SETI endeavors in the decades to come.

The legacy of early SETI initiatives endures as a testament to human curiosity and the enduring quest to explore the cosmic frontier. While the search for extraterrestrial intelligence continues to evolve, these pioneering efforts remain integral to the broader scientific endeavor of unraveling the mysteries of the universe and seeking potential cosmic neighbors.

5.4.2. SETI's Journey and Future

The journey of the Search for Extraterrestrial Intelligence (SETI) has been marked by scientific curiosity, technological innovation, and the unwavering pursuit of answers to one of humanity's most profound questions: Secrets of the Cosmos As we reflect on SETI's past and peer into its future, we find a tale of resilience and hope.

A Quest That Endures: SETI's journey began with pioneering initiatives in the mid-20th century and has persisted for over six decades. It is a quest that endures, driven by the innate human desire to explore the unknown and connect with potential cosmic neighbors.

Advancements in Technology: Over the years, SETI has benefited from rapid advancements in technology. The digital revolution transformed signal processing, allowing scientists to analyze vast datasets with unprecedented precision. Today, supercomputers and sophisticated algorithms aid in the search for elusive extraterrestrial signals.

The Breakthrough Listen Initiative: In recent years, the Breakthrough Listen Initiative has emerged as a prominent player in the field of SETI. Supported by visionary investors, this initiative employs some of the world's most powerful telescopes and cutting-edge instrumentation to scan the skies for signs of intelligent life.

The Role of Citizen Scientists: Citizen scientists, too, play an integral role in SETI. Initiatives like SETI@home have engaged millions of volunteers in processing radio signal data using their personal computers, turning the search into a global collaborative effort.

New Horizons: SETI's future is marked by expanding horizons. Upcoming observatories like the Square Kilometre Array (SKA) promise to enhance our capability to explore cosmic radio signals. The James Webb Space Telescope (JWST) will probe exoplanet atmospheres, furthering the quest for biosignatures.

The Search for Technosignatures: While traditional SETI efforts focus on radio signals, the search for technosignatures has gained traction. This approach seeks evidence of advanced technologies or engineering feats that could be indicative of intelligent civilizations.

Ethical and Philosophical Considerations: As SETI advances, so do ethical and philosophical considerations. The potential implications of contact with extraterrestrial civilizations spark discussions on the responsible handling of such discoveries and their societal impacts.

The Perseverance of Curiosity: SETI's journey embodies the perseverance of human curiosity. It is a testament to our willingness to explore the cosmos, to push boundaries, and to seek answers to questions that transcend borders and cultures.

As SETI continues its voyage into the cosmic unknown, it serves as a reminder that our quest to connect with the universe is an enduring endeavor. The search for extraterrestrial intelligence is not solely a scientific pursuit; it is a testament to the human spirit of exploration and our shared aspiration to unlock the mysteries of the cosmos.

5.4.3. At the Frontiers of Science

At the intersection of science, technology, and human curiosity, the Search for Extraterrestrial Intelligence (SETI) stands as a beacon guiding us to the frontiers of knowledge and the very limits of our understanding.

Pushing the Technological Envelope: SETI has always been at the forefront of technological innovation. It challenges us to develop increasingly sophisticated instruments and computing capabilities. Breakthroughs in signal processing, artificial intelligence, and data analysis are direct outcomes of the quest to decipher potential cosmic messages.

The Role of AI: Artificial intelligence (AI) plays a pivotal role in contemporary SETI. Machine learning algorithms sift through vast datasets, identifying patterns that human observers might miss. AI-driven techniques enable us to search for complex and subtle signals amidst the cosmic noise.

The Fermi Paradox: SETI confronts the enigma known as the Fermi Paradox—given the vast number of potentially habitable planets in the universe, why haven't we encountered extraterrestrial civilizations? This paradox fuels ongoing debates and inspires creative solutions, from the "zoo hypothesis" to the possibility of "great filters."

Astrobiology and Biosignatures: The study of astrobiology, closely intertwined with SETI, investigates the conditions necessary for life to emerge and thrive. Scientists explore exoplanets for biosignatures, indicators of life, such as specific chemical imbalances or atmospheric compositions.

The Search for Technological Signatures: Beyond radio signals, SETI increasingly focuses on detecting technosignatures—indications of advanced technology. This includes the hunt for megastructures, energy sources, or other artifacts that advanced civilizations might produce.

Interstellar Missions: The possibility of interstellar missions to explore nearby star systems is now within our technological reach. Projects like Breakthrough Starshot envision sending tiny spacecraft propelled by lasers on journeys to our nearest stellar neighbors.

The Philosophical Quest: SETI delves into the realms of philosophy, ethics, and societal implications. The potential discovery of extraterrestrial intelligence raises profound questions about our place in the universe and how we should responsibly engage with other civilizations.

Educating and Inspiring: SETI serves as a powerful tool for science education and public engagement. It inspires generations of scientists, engineers, and dreamers to pursue careers in space science and exploration.

The Future Awaits: As SETI advances, the future holds both uncertainty and boundless potential. It is a journey into the unknown, a quest that fuels our imagination, and a testament to human curiosity and the indomitable spirit of exploration.

SETI represents our shared quest to reach the frontiers of science, to understand the cosmos, and to forge connections beyond our planet. It serves as a testament to the enduring human spirit that propels us to explore the mysteries of the universe, guided by the hope of one day making contact with intelligent beings from the stars.

CHAPTER 6

What Can We Expect in the Near Future?

6.1. Technological Advancements and Space Exploration

6.1.1. Advanced Technology and the Limits of Space

The pursuit of space exploration has continually pushed the boundaries of technology, engineering, and human understanding. As we delve into the complexities of the cosmos, we are met with both the promise of advanced technology and the daunting limits that the vast expanse of space imposes upon our endeavors.

Spacecraft Propulsion Innovations: Advanced propulsion technologies are at the forefront of space exploration. From chemical rockets to ion propulsion and beyond, engineers are tasked with developing systems capable of propelling spacecraft at unprecedented speeds. The challenge lies in achieving efficient propulsion methods that can carry missions to distant celestial bodies within reasonable timeframes.

Interstellar Travel Aspirations: Humanity's quest for interstellar travel is marked by the dream of reaching neighboring star systems within a human lifetime. Concepts like nuclear propulsion and laser propulsion are explored, offering the potential to propel spacecraft at a significant fraction of the speed of light. However, the immense energy requirements and engineering hurdles remain substantial challenges.

Spacecraft Autonomy and Artificial Intelligence: The distances involved in interstellar travel necessitate spacecraft autonomy. Artificial intelligence (AI) systems play a pivotal role in navigating and managing complex missions. AI-driven spacecraft can adapt to unforeseen circumstances, execute scientific tasks, and even make real-time decisions.

Communicating Across Cosmic Distances: The vastness of space presents formidable challenges in communication. Delays in signal transmission become significant over interstellar distances, necessitating advanced communication technologies. Laser-based communication and autonomous relay networks are envisioned to bridge the gap between distant spacecraft and Earth.

Radiation Protection and Health: Extended space missions, especially those beyond our solar system, must address radiation hazards. Developing advanced shielding materials and medical countermeasures is essential to safeguarding astronauts on long-duration missions.

Resource Utilization Beyond Earth: To sustain missions over extended periods, resource utilization beyond Earth's boundaries becomes imperative. Technologies for in-situ resource utilization (ISRU) enable astronauts to extract vital resources, such as water and oxygen, from extraterrestrial environments.

Life Support Systems: Developing closed-loop life support systems that can efficiently recycle resources like water and air is crucial for self-sustaining missions. These systems reduce the dependence on resupply from Earth, extending mission durations.

The Enigma of Faster-Than-Light Travel: Concepts like warp drives and wormholes, popularized in science fiction, represent the desire to overcome the limitations of space and time. However, these ideas remain speculative and entail theoretical challenges, including the need for exotic matter or energy.

Human-Computer Integration: As missions extend further into the cosmos, the integration of human cognition with advanced computing becomes paramount. Brain-computer interfaces and augmented reality systems can enhance astronaut capabilities and decision-making.

Ethical Considerations: With the potential for interstellar travel on the horizon, ethical dilemmas emerge. These include questions about the preservation of extraterrestrial ecosystems and the responsible exploration of potentially inhabited exoplanets.

The Limits of Physics: Finally, the limits imposed by the laws of physics themselves present perhaps the most daunting challenge. The speed of light, as an ultimate cosmic speed limit, sets constraints on the feasibility of interstellar travel and communication.

As humanity's technological prowess advances, our quest to reach the stars reveals both the remarkable innovations of advanced technology and the inherent challenges imposed by the cosmos. It is a journey that compels us to push the limits of our understanding, confront the unknown, and continually strive to bridge the gap between our terrestrial home and the boundless expanse of space.

6.1.2. Future Spacecraft

The future of space exploration promises a new generation of spacecraft, equipped with cutting-edge technologies and innovative designs that will redefine our capabilities beyond Earth's orbit. These spacecraft represent the vanguard of human exploration in the cosmic theater.

Advanced Propulsion Technologies: Future spacecraft will harness advanced propulsion technologies to enable missions to distant celestial destinations. Concepts like nuclear thermal propulsion, solar sails, and even antimatter propulsion are under consideration. These innovations aim to reduce travel times and expand the reach of human exploration.

Interstellar Probes: Ambitious interstellar missions are on the horizon, with the goal of sending spacecraft to neighboring star systems. These probes will need to endure the rigors of space for centuries, relying on advanced materials and autonomous systems for navigation and communication.

In-Situ Resource Utilization: To sustain long-duration missions, future spacecraft will incorporate in-situ resource utilization (ISRU) capabilities. This means extracting essential resources from extraterrestrial environments, such as water ice on the Moon or Mars, to support human activities.

Autonomous Systems and AI: Spacecraft of the future will rely heavily on autonomous systems and artificial intelligence (AI). These AI-driven systems will manage complex mission operations, perform scientific analyses, and adapt to unforeseen challenges, reducing the need for constant human intervention.

Laser-Based Communication: The vast distances in space demand innovative communication solutions. Laser-based communication offers higher data transfer rates and lower signal degradation over interplanetary and interstellar distances, revolutionizing how spacecraft communicate with Earth.

Bio-Inspired Designs: Future spacecraft may draw inspiration from biology, utilizing designs inspired by nature to enhance efficiency and adaptability. Biomimetic technologies could lead to spacecraft that can self-repair, adapt to changing conditions, and perform intricate tasks.

Nanotechnology and Miniaturization: Advances in nanotechnology and miniaturization enable the development of small, highly capable spacecraft. CubeSats and microprobes can explore distant targets, such as asteroids or the outer solar system, at a fraction of the cost of traditional missions.

Human-Centered Design: Spacecraft designed for human missions beyond Earth will prioritize human factors, including comfort, safety, and psychological well-being. Habitats in space will mimic Earth's conditions as closely as possible to ensure the health and productivity of astronauts.

Space Tourism Vehicles: The burgeoning space tourism industry will see the development of spacecraft designed to carry paying passengers on suborbital and orbital flights. These vehicles will prioritize safety, comfort, and the thrill of space exploration for private individuals.

Sustainable Spacecraft: Sustainability is a key consideration for future spacecraft. Sustainable propulsion systems, recycling technologies, and responsible disposal measures will be integrated into spacecraft design to minimize the environmental impact of space exploration.

International Collaboration: Future spacecraft projects are likely to involve international collaboration, pooling expertise and resources to tackle ambitious missions. Collaborative efforts aim to extend humanity's reach further into the cosmos and foster global cooperation in space exploration.

The spacecraft of tomorrow are poised to propel us deeper into the cosmos, transforming our understanding of the universe and our place within it. With their advanced technologies and innovative designs, these spacecraft embody the spirit of human exploration and our unceasing quest to explore the cosmic frontier.

6.1.3. The Role of Humans in the Future

As we chart a course into the future of space exploration, the role of humans remains central to our endeavors beyond Earth. While robotic missions and autonomous systems pave the way, human presence and engagement in space exploration continue to hold immense significance.

The Pioneering Spirit: Humans have always been explorers, driven by an innate curiosity to venture into the unknown. The future of space exploration is no exception, as humans eagerly embrace the challenges and rewards of exploring the cosmos.

Interplanetary Colonization: Ambitious plans to establish human colonies on celestial bodies such as the Moon and Mars are in the works. These colonies represent stepping stones towards becoming a multi-planetary species, ensuring the survival and growth of humanity beyond Earth.

Human-Technology Synergy: The synergy between humans and technology is at the heart of future space exploration. Human expertise, adaptability, and problem-solving skills complement the capabilities of advanced spacecraft and robotic systems.

Long-Duration Space Missions: Future missions will require astronauts to embark on long-duration journeys, pushing the boundaries of human endurance. Innovations in life support systems, medical care, and psychological support are essential to sustaining human health and well-being in space.

Human-Centered Design: Spaceships and habitats designed for human presence prioritize comfort, safety, and psychological well-being. Habitats aim to replicate Earth-like conditions, fostering a sense of home in the harsh environments of space.

Exploring the Unknown: Humans bring a unique capacity for exploration, intuition, and adaptability. Astronauts on distant missions will have the ability to make real-time decisions, investigate unexpected discoveries, and respond to unforeseen challenges.

Scientific Discovery: Human scientists and researchers will conduct experiments and observations in space that are simply not possible with autonomous systems alone. Their presence enables real-time scientific investigations, expanding our understanding of the cosmos.

Space Tourism: As space tourism gains momentum, more individuals will have the opportunity to experience space firsthand. These private space travelers will contribute to the democratization of space, fostering a broader interest in space exploration.

International Collaboration: Collaborative efforts in space exploration will transcend national boundaries. International cooperation on space missions will strengthen our collective knowledge and capabilities, promoting peaceful exploration of the cosmos.

Inspiration and Education: Human presence in space continues to inspire future generations and serves as a powerful tool for science education. Astronauts become ambassadors of science and exploration, igniting the passion for space in the hearts of young minds.

The Ethical Frontier: As humans venture deeper into space, ethical considerations gain prominence. The responsible and sustainable exploration of space, preservation of extraterrestrial environments, and equitable sharing of cosmic resources become pressing issues.

The role of humans in the future of space exploration is multifaceted, encompassing scientific discovery, pioneering spirit, international collaboration, and the pursuit of knowledge. As we embark on this cosmic journey, humans stand as both explorers and stewards of the universe, carrying with them the dreams and aspirations of a species destined to reach the stars.

6.2. Humanity's Quest for Extraterrestrial Life

6.2.1. Humanity's Progression

The progression of humanity in the realm of space exploration is a testament to our relentless pursuit of knowledge, adventure, and the desire to expand the boundaries of our existence. As we look to the future, it is crucial to reflect on the remarkable journey that has brought us to this point.

The Dawn of Space Exploration: The journey began with the launch of the first artificial satellite, Sputnik 1, by the Soviet Union in 1957. This historic event marked the dawn of the space age and ignited the space race between superpowers.

The Apollo Era: The United States took a giant leap forward with the Apollo program, culminating in the historic Apollo 11 moon landing in 1969. This achievement demonstrated humanity's ability to reach celestial bodies beyond Earth.

Space Shuttle Era: The development of the Space Shuttle program in the 1970s ushered in an era of reusable spacecraft. These shuttles enabled regular human access to space and facilitated groundbreaking missions such as the Hubble Space Telescope deployment.

International Space Station (ISS): The construction and operation of the ISS, initiated in the late 1990s, symbolized international collaboration in space exploration. The station serves as a platform for scientific research and international cooperation in microgravity.

Robotic Exploration: Robotic missions to Mars, Jupiter's moons, and beyond have expanded our understanding of the solar system. Rovers like Curiosity and Perseverance continue to explore the Martian surface, while spacecraft like Juno probe the mysteries of distant gas giants.

Commercial Spaceflight: The emergence of private space companies, such as SpaceX, Blue Origin, and Virgin Galactic, has transformed space access. These companies aim to make space more accessible, from launching satellites to enabling space tourism.

Humanity Beyond Low Earth Orbit: As we venture beyond low Earth orbit, plans to return humans to the Moon through programs like Artemis and establish a presence on Mars gain momentum. These endeavors represent the next chapters in our cosmic story.

Emerging Space Nations: An increasing number of nations are entering the space arena, contributing to the global landscape of space exploration. Spacefaring nations collaborate on missions, share knowledge, and advance the peaceful exploration of space.

Scientific Discoveries: Space missions have revealed the mysteries of the cosmos, from the composition of distant planets to the nature of black holes. These discoveries reshape our understanding of the universe and ignite scientific curiosity.

Challenges and Opportunities: Humanity faces challenges in the realms of space debris management, sustainability, and ethical considerations. However, these challenges also offer opportunities for innovation, international cooperation, and responsible exploration.

Educating Future Generations: Space exploration continues to inspire and educate future generations. It fosters interest in science, technology, engineering, and mathematics (STEM) fields and encourages the pursuit of careers in space science and exploration.

In the grand tapestry of space exploration, humanity's progression represents a remarkable chapter—one characterized by ambition, collaboration, and the relentless pursuit of the unknown. As we look ahead to the future, we carry with us the knowledge and experiences gained from our cosmic journey, poised to embark on new adventures that will shape the destiny of our species.

6.2.2. The Pursuit of Science and Exploration

The pursuit of science and exploration has been the driving force behind humanity's journey into the cosmos. From the earliest observations of the night sky to the sophisticated missions of today, our quest for knowledge and understanding has illuminated the mysteries of the universe.

Stellar Observations of Antiquity: The ancient civilizations of Mesopotamia, Egypt, Greece, and China observed the heavens, laying the foundations for astronomy. Their discoveries included the motions of planets, the phases of the Moon, and the recognition of constellations.

The Renaissance and the Telescope: The Renaissance era saw the refinement of optical instruments, leading to the invention of the telescope. Pioneers like Galileo Galilei used telescopes to make groundbreaking observations, including the moons of Jupiter and the phases of Venus, challenging existing astronomical models.

Advances in Astrophysics: The 19th and early 20th centuries witnessed significant advances in astrophysics. Innovations like spectroscopy allowed scientists to analyze the composition of stars and distant galaxies, unraveling the secrets of the cosmos.

Space Age Beginnings: The launch of the first artificial satellite, Sputnik 1, in 1957 marked the beginning of the space age. The subsequent launch of human-crewed missions and robotic spacecraft opened new frontiers for scientific exploration.

Lunar Exploration: The Apollo program, with its historic moon landings, expanded our understanding of lunar geology and planetary formation. Samples brought back from the Moon provided invaluable insights into the history of our solar system.

Planetary Probes: Robotic missions to other planets, such as the Viking landers on Mars and the Voyager probes exploring the outer solar system, delivered a wealth of data about our celestial neighbors, their atmospheres, and geologies.

Space Telescopes: The launch of space telescopes like Hubble, Chandra, and Kepler revolutionized our view of the universe. These observatories provided breathtaking images and discoveries related to distant galaxies, black holes, and exoplanets.

Exploring the Inner Solar System: Missions to Mercury, Venus, and asteroids have deepened our understanding of these worlds and planetary processes.

Rovers like Curiosity continue to explore the Martian surface.

Outer Solar System and Beyond: Spacecraft like New Horizons have ventured into the distant reaches of the solar system, offering close-up views of Pluto and the Kuiper Belt. The Voyager probes have even entered interstellar space, carrying messages from Earth.

Astrobiology and the Search for Life: The study of extremophiles on Earth and the discovery of potentially habitable exoplanets have fueled the field of astrobiology. Scientists are actively searching for signs of life beyond our planet.

Cosmic Mysteries: Space exploration has illuminated mysteries such as the nature of dark matter and dark energy, the origins of cosmic structures, and the behavior of black holes. These puzzles continue to drive scientific inquiry.

Collaboration and Knowledge Sharing: International cooperation in space missions has become the norm, with nations pooling resources and knowledge for mutual benefit. This spirit of collaboration has expanded our global perspective on space exploration.

Educational Outreach: Space agencies and organizations worldwide engage in educational outreach, inspiring future generations of scientists, engineers, and explorers to continue the pursuit of knowledge in the cosmos.

The pursuit of science and exploration remains an enduring testament to human ingenuity and curiosity. From ancient stargazers to modern astronomers, our collective journey to understand the universe represents a noble endeavor that transcends time and borders, propelling us ever further into the cosmic unknown.

6.2.3. The Pioneers of the Future

As humanity advances further into the cosmos, a new generation of pioneers emerges to lead the way. These pioneers are individuals, organizations, and nations dedicated to shaping the future of space exploration and ensuring that humanity's presence in the cosmos continues to grow.

Visionary Space Entrepreneurs: Visionary entrepreneurs like Elon Musk, Jeff Bezos, and Richard Branson have established private space companies such as

SpaceX, Blue Origin, and Virgin Galactic. They aim to make space more accessible, reduce launch costs, and enable commercial space travel.

International Collaborators: International cooperation in space exploration has never been stronger. Collaborative efforts between space agencies like NASA, ESA, Roscosmos, CNSA, and ISRO demonstrate the power of unity in achieving ambitious cosmic goals.

Inspiring Astronauts: Astronauts continue to serve as role models and sources of inspiration for future generations. Their courage, dedication, and scientific contributions remain central to human space exploration.

Space Scientists and Researchers: Scientists and researchers across the globe work tirelessly to unlock the mysteries of the universe. Their work includes studying exoplanets, cosmic phenomena, and the potential for life beyond Earth.

Next-Generation Engineers: Engineers and innovators are designing the spacecraft, technologies, and systems that will enable future missions. Their contributions span propulsion, spacecraft design, artificial intelligence, and more.

Educational Advocates: Educational programs and advocates promote STEM (science, technology, engineering, and mathematics) fields, inspiring young minds to pursue careers in space science and exploration.

Robotic Explorers: Robotic missions and autonomous systems continue to advance our understanding of the cosmos. Rovers, landers, and probes investigate distant planets, asteroids, and comets, laying the groundwork for future human missions.

Space Policy and Regulation: Policymakers and regulators shape the legal framework for space activities, ensuring responsible and sustainable exploration, resource utilization, and international cooperation.

Space Tourism Pioneers: The emerging space tourism industry is led by pioneers who aim to make suborbital and orbital space travel accessible to private individuals, ushering in a new era of civilian spaceflight.

Astrobiologists: Astrobiologists explore extreme environments on Earth and search for signs of life beyond our planet. Their work drives the quest to find habitable worlds and answer the question of whether life exists elsewhere in the universe.

Space Visionaries: Visionaries across various disciplines, including science fiction writers, artists, and futurists, inspire us to dream of what lies beyond and envision a future where humanity thrives in the cosmos.

Ethical Thinkers: Ethical thinkers and scholars engage in discussions about the moral and ethical considerations of space exploration, including issues related to resource utilization, preservation of celestial bodies, and international cooperation.

These pioneers of the future are united by a common goal: to propel humanity further into the cosmos, unlocking the secrets of the universe, and ensuring that our species continues to explore, adapt, and thrive in the vast cosmic frontier. Their contributions pave the way for a future where the possibilities of space exploration are limited only by our imagination and determination.

6.3. Future Communication Techniques and Strategies

6.3.1. Advancements in Communication

Communication in the realm of space exploration has undergone remarkable advancements, transforming the way we exchange information and connect with spacecraft and astronauts in the cosmos. These advancements are essential for enabling effective space missions, scientific discoveries, and human presence beyond Earth.

Reliable Data Transmission: Advancements in communication technology have led to highly reliable data transmission between Earth and spacecraft. Improved signal processing and error correction techniques ensure that data arrives intact, even from the far reaches of the solar system.

Deep Space Network: NASA's Deep Space Network (DSN) plays a pivotal role in space communication. This network of large antennas located around the world facilitates continuous contact with spacecraft, enabling real-time command and data retrieval.

Laser Communication: Laser communication, or optical communication, offers higher data transfer rates than traditional radio waves. NASA's Laser Communications Relay Demonstration (LCRD) and other projects explore the potential of laser communication for future missions.

Interplanetary Internet: The concept of an interplanetary internet is emerging, allowing for seamless communication between Earth, spacecraft, and eventually, human colonies on other celestial bodies. This technology will enable data sharing and remote control of equipment on distant planets.

Autonomous Communication: As missions venture deeper into space, autonomous communication becomes crucial. Autonomous systems can adapt to changing conditions and delays in communication, ensuring the safety and success of missions.

Quantum Communication: Quantum communication promises secure and unhackable communication channels, a vital consideration for transmitting sensitive data in space. Quantum key distribution experiments are being conducted to test the feasibility of this technology.

Real-Time Video and Audio: Advancements in communication have enabled real-time video and audio exchanges between astronauts aboard the International Space Station (ISS) and mission control on Earth. This enhances collaboration, troubleshooting, and morale during long missions.

Astronomical Observations: Radio telescopes and observatories communicate with spacecraft to coordinate observations of distant celestial objects. These collaborations expand our understanding of the universe and uncover cosmic phenomena.

Earth-Mars Communication: The challenge of communication between Earth and Mars, with varying distances and orbital positions, is met with sophisticated relay systems like NASA's Mars Reconnaissance Orbiter (MRO) and Mars Odyssey, ensuring continuous connectivity.

International Collaboration: International space agencies collaborate on communication infrastructure, sharing resources and expertise to maintain contact with spacecraft beyond Earth's orbit. Such collaboration strengthens global capabilities in space exploration.

Private Sector Initiatives: Private companies are investing in advanced communication systems for their space missions, contributing to the evolution of space communication technology and infrastructure.

Public Engagement: Improved communication technologies allow the public to engage with space missions through live broadcasts, social media, and interactive educational initiatives, fostering greater interest in space exploration.

The advancements in space communication are not only vital for mission success but also for enabling humanity's continued presence in the cosmos. As we pioneer deeper into space, these innovations in communication will ensure that we remain connected to the mysteries of the universe and to each other, transcending the boundaries of our home planet.

6.3.2. Communicating with Unknown Entities

As humanity explores the cosmos and contemplates the existence of extraterrestrial life, the challenge of communicating with unknown entities becomes a central consideration. This endeavor requires careful planning, ethical considerations, and a profound understanding of potential contact scenarios.

Interstellar Messaging: Initiatives like the "Breakthrough Message" project seek to create interstellar messages that may one day be sent into the cosmos. These messages aim to represent humanity and our shared values while potentially reaching distant intelligent civilizations.

Messaging Content: Crafting the content of interstellar messages involves choosing universal symbols, mathematical notations, and representations of basic science. The challenge is to create messages that can be understood by beings with no prior knowledge of human languages or culture.

Cultural and Ethical Considerations: Communicating with unknown entities raises complex ethical questions. How do we ensure that our messages are culturally sensitive and non-threatening? What safeguards should be in place to protect Earth and its inhabitants?

Messaging as a Two-Way Process: Communication with extraterrestrial civilizations may not be a one-sided endeavor. We must be prepared for the possibility of receiving messages from unknown entities and interpreting their content.

Language of Mathematics: Mathematics is often considered a universal language that can transcend linguistic barriers. Mathematical principles, formulas, and concepts may serve as a common ground for communication.

Active versus Passive Messaging: Some argue for a cautious approach, advocating passive messaging that minimizes the risk of revealing our existence. Others support active messaging, actively transmitting signals into space in hopes of initiating contact.

Messaging Protocols: The development of standardized messaging protocols is essential to ensure clear and consistent communication with potential extraterrestrial intelligences. Protocols would dictate how information is structured and transmitted.

Message Detection and Analysis: On Earth, the search for extraterrestrial intelligence (SETI) involves the detection and analysis of potential signals from space. Sophisticated algorithms and technology are employed to identify patterns and anomalies in incoming data.

Communication Barriers: Communicating with unknown entities requires overcoming vast distances, signal degradation over interstellar travel, and the inherent limitations of the speed of light. These challenges necessitate innovative solutions and strategies.

First Contact Scenarios: Scientists, policymakers, and ethicists contemplate the implications of first contact scenarios. How should we respond if we receive a signal? What are the consequences of revealing ourselves to advanced civilizations?

International Frameworks: International agreements and frameworks are necessary to address the global nature of potential contact with extraterrestrial civilizations. These agreements would guide humanity's response and ensure a coordinated approach.

Ultimately, the endeavor to communicate with unknown entities is a testament to human curiosity, imagination, and our desire to reach out to the cosmos. As we navigate the complexities of interstellar messaging, we must remain mindful of the profound impact such contact could have on our worldview, ethics, and the future of our species.

6.3.3. The Codes of Communication in the Universe

In the quest to communicate with potential extraterrestrial intelligences, scientists and researchers must grapple with the challenge of decoding the messages of the cosmos. This challenge extends beyond linguistic and cultural barriers and involves understanding the fundamental codes of communication that may exist in the universe.

Universal Language of Physics: Physics is often regarded as the universal language of the cosmos. Fundamental constants, physical laws, and mathematical principles could serve as a common ground for communication, transcending the need for language.

Mathematics as a Bridge: Mathematics, with its precision and universality, may provide a key to decoding extraterrestrial messages. Mathematical concepts such as prime numbers and mathematical constants like pi could be used as a basis for interstellar communication.

Patterns and Symmetry: The universe is replete with patterns, symmetry, and fractal structures. These recurring motifs may carry meaning and could be a form of cosmic communication. Recognizing these patterns is essential for decoding potential messages.

The Language of Chemistry: Chemistry, the science of matter and its interactions, may hold clues to decoding alien messages. The arrangement of atoms and molecules could convey information about the composition and chemistry of extraterrestrial environments.

Binary Code: Binary code, consisting of ones and zeros, is a fundamental language in digital communication. Alien civilizations might employ binary code or similar binary-like systems in their messages, which we could decipher through pattern recognition.

Prime Numbers as a Signature: Prime numbers, those divisible only by one and themselves, are considered a mathematical signature of intelligence. Efforts to decipher potential messages may involve searching for prime number sequences.

Hierarchical Structures: Just as human languages have hierarchical structures, extraterrestrial messages might exhibit similar patterns. Decoding efforts could involve identifying hierarchies in information presentation.

Linguistic Approaches: Linguists and language experts play a critical role in decoding potential extraterrestrial messages. Analyzing linguistic features, syntax, and semantics may provide insights into the structure of alien languages.

Machine Learning and AI: Advanced machine learning and artificial intelligence algorithms are employed to identify and decode patterns in vast datasets. These technologies can assist in the analysis of complex messages from the cosmos.

Interdisciplinary Collaboration: The challenge of decoding cosmic messages requires interdisciplinary collaboration, bringing together experts in physics, mathematics, linguistics, computer science, and other fields to collectively tackle the problem.

Ethical Considerations: As we delve into the realm of decoding alien messages, ethical considerations arise. Researchers must address issues related to respecting the privacy and intentions of potential extraterrestrial civilizations.

Continual Vigilance: The search for cosmic messages is an ongoing endeavor, necessitating continual vigilance and analysis of data from telescopes, space missions, and SETI initiatives.

The codes of communication in the universe are a profound puzzle waiting to be deciphered. As we delve deeper into this quest, we must remain open to the possibility of decoding messages that could expand our understanding of the cosmos and potentially reveal insights into the nature of intelligent life beyond Earth.

6.4. One Step Closer to the Mysteries of the Universe

6.4.1. From the Unknown to the Known

As humanity embarks on a journey into the unknown reaches of the cosmos, we are met with the tantalizing prospect of transforming the unknown into the known. This profound transformation involves the acquisition of knowledge, the unraveling of cosmic mysteries, and the expansion of our understanding of the universe.

Unveiling Cosmic Mysteries: Exploring distant celestial bodies, probing the depths of space, and conducting in-depth observations allow us to unveil long-standing cosmic mysteries. We seek to understand the origins of the universe, the nature of dark matter and dark energy, and the existence of extraterrestrial life.

Mapping Uncharted Territories: The exploration of space involves mapping uncharted territories, from the surfaces of distant planets and moons to the farthest reaches of the cosmos. These maps serve as valuable resources for future missions and scientific endeavors.

Searching for Habitable Worlds: The quest for habitable worlds beyond Earth is a central focus of space exploration. Scientists search for exoplanets with the potential to support life, with a keen eye on those within the habitable zone of distant stars.

Investigating Cosmic Phenomena: Space missions and observatories provide opportunities to investigate cosmic phenomena, such as supernovae, black holes, pulsars, and quasars. These studies shed light on the fundamental processes shaping the universe.

Studying Extraterrestrial Life: The search for extraterrestrial life involves studying extremophiles on Earth and exploring the potential for life on other celestial bodies. This field of astrobiology opens the door to profound discoveries.

Understanding the Cosmos: The cosmos is a vast and interconnected system. Space exploration allows us to grasp the interplay between celestial bodies, galaxies, and the larger cosmic web, ultimately deepening our understanding of the universe's structure and evolution.

Expanding Human Presence: As humans venture beyond Earth, we expand our presence in the cosmos. The establishment of research stations on the Moon, Mars, and beyond offers opportunities for scientific exploration and potential colonization.

Pushing Technological Boundaries: Space exploration drives technological advancements, from propulsion systems and robotics to life support and communication technologies. These innovations have far-reaching implications for industries on Earth.

Inspiring Future Generations: Space exploration captivates the imagination and inspires future generations of scientists, engineers, and explorers. It fosters a spirit of curiosity and innovation that extends beyond our planet.

International Cooperation: Collaborative efforts in space exploration unite nations and promote peaceful cooperation. International partnerships enable the pooling of resources, expertise, and knowledge for ambitious space missions.

Preserving the Cosmic Environment: As we venture into space, we must act as stewards of the cosmic environment. Ethical considerations include the responsible use of celestial resources and the preservation of pristine cosmic environments.

The transformation from the unknown to the known is an ongoing journey that transcends borders and spans generations. With each new discovery, humanity takes a step closer to unraveling the mysteries of the cosmos, fulfilling our innate curiosity, and forging a path toward a future where the boundaries of the unknown continue to recede.

6.4.2. The Horizons of Science

As humanity's exploration of the cosmos advances, the horizons of science expand in tandem. The frontiers of knowledge push further into the unknown, unveiling new insights, raising profound questions, and redefining our understanding of the universe. Within this ever-evolving landscape, science plays a pivotal role in shaping the future of space exploration.

Unraveling the Nature of Dark Matter: One of the most pressing questions in cosmology is the nature of dark matter—a mysterious substance that comprises a significant portion of the universe's mass. Ongoing research and experiments seek to identify and understand this elusive component.

Probing the Secrets of Dark Energy: Dark energy, responsible for the accelerating expansion of the universe, remains a profound enigma. Scientists employ astronomical observations, such as those from the Hubble Space Telescope, to delve deeper into the nature of this enigmatic force.

Exploring Quantum Gravity: The unification of general relativity and quantum mechanics to form a theory of quantum gravity represents a paramount challenge in theoretical physics. Advancements in this field could revolutionize our understanding of the fundamental forces governing the universe.

Investigating the Multiverse Hypothesis: The concept of a multiverse—a vast ensemble of parallel universes—captivates the imagination of physicists and cosmologists. Research explores the possibility of multiple universes with differing physical laws.

Advancing Astrobiology: Astrobiology continues to expand the search for life beyond Earth. Investigations into extremophiles, subsurface oceans, and potential biosignatures on exoplanets drive our understanding of the conditions necessary for life to flourish.

Studying Exoplanets: The discovery of thousands of exoplanets has transformed our perspective on planetary systems. Ongoing missions and telescope observations provide insights into the atmospheres and habitability of distant worlds.

Unlocking Quantum Computing: The development of quantum computing holds the promise of solving complex problems far more efficiently than classical computers. Quantum algorithms could revolutionize data analysis and simulations in space exploration.

Harnessing Nuclear Fusion: Nuclear fusion, the process powering stars, remains an elusive energy source on Earth. Advancements in fusion research bring us closer to achieving sustainable and clean energy production.

Space Medicine and Long-Duration Missions: Space medicine and research on the effects of long-duration spaceflight on the human body are vital for future deep space exploration. These studies inform strategies for astronaut health and well-being.

Artificial Intelligence in Space: Artificial intelligence and machine learning play critical roles in autonomous spacecraft operation, data analysis, and mission planning. AI enhances the efficiency and capabilities of space missions.

Emerging Propulsion Technologies: Advanced propulsion technologies, such as ion propulsion and solar sails, enable faster and more efficient space travel. These innovations reduce travel time and open up new possibilities for exploration.

Ethical Considerations in Space Science: As we venture further into the cosmos, ethical considerations become increasingly important. Scientists and policymakers grapple with issues related to planetary protection, resource utilization, and responsible exploration.

Infinite Curiosity: The horizons of science expand infinitely alongside human curiosity. The pursuit of knowledge drives innovation, inspires exploration, and fosters a deeper connection between humanity and the vast cosmos.

The horizons of science continue to beckon, inviting us to embark on a journey of discovery and enlightenment. As we traverse these new frontiers, we not only deepen our understanding of the universe but also redefine the boundaries of what is possible, pushing the limits of human knowledge and imagination.

6.4.3. Questions of the Future

As humanity ventures further into the cosmos, a myriad of profound questions about the future of space exploration and our place in the universe emerge. These questions encompass scientific, ethical, and existential dimensions, challenging us to contemplate the vast expanse of possibilities that lie ahead.

The Destiny of Humanity: What destiny awaits humanity as we expand our presence in the cosmos? Will we establish permanent settlements on distant celestial bodies, becoming a multi-planetary species?

The Search for Extraterrestrial Life: Will we make the historic discovery of extraterrestrial life, reshaping our understanding of biology and the potential for life beyond Earth?

Interstellar Travel: Can humanity overcome the vast distances of interstellar space to embark on journeys to neighboring star systems? What propulsion technologies and strategies will make this endeavor possible?

The Nature of Dark Matter and Dark Energy: Will we unlock the mysteries of dark matter and dark energy, gaining insight into the fundamental forces shaping the universe?

The Multiverse Hypothesis: Is the multiverse hypothesis a valid explanation for the cosmos? If so, what implications does it hold for our understanding of reality?

Quantum Gravity and Unified Theories: Can we successfully unify the theories of general relativity and quantum mechanics to achieve a comprehensive understanding of the fundamental forces of nature?

Artificial Intelligence and Space Exploration: How will artificial intelligence continue to revolutionize space exploration, from autonomous spacecraft to data analysis and mission planning?

Sustainability in Space: What strategies and technologies will ensure sustainable and responsible exploration of celestial bodies while preserving their pristine environments?

The Ethics of Contact: How will humanity navigate the ethical challenges of potential contact with extraterrestrial civilizations, respecting their autonomy and intentions?

The Role of International Cooperation: Will international collaboration expand, enabling humanity to address the global nature of space exploration and its associated challenges?

Innovations in Propulsion: What groundbreaking propulsion technologies will emerge to propel spacecraft deeper into the cosmos, reducing travel time and enabling ambitious missions?

Space Mining and Resource Utilization: How will space mining and the utilization of extraterrestrial resources transform Earth's economy and industry?

Long-Term Survival: What measures will be taken to ensure the long-term survival of humanity, including planetary protection and strategies for adapting to cosmic environments?

These questions of the future are a testament to the boundless curiosity, innovation, and ambition of humanity. As we strive to answer these inquiries, we embark on a journey that extends beyond our lifetimes, shaping the destiny of generations to come and defining our enduring legacy in the cosmos.

Conclusion: Secrets of the Cosmos

Exploring the Cosmos: A Journey Beyond Imagination

In the relentless quest to unravel the mysteries of the universe and explore the potential existence of extraterrestrial life, humanity has embarked on an extraordinary odyssey that transcends the boundaries of our planet. This monumental journey is a testament to our insatiable curiosity, boundless innovation, and unwavering determination.

Our expedition commences with a profound exploration of the birth and expansion of the universe itself. At the very genesis of time, we cast our gaze back to the epochal moment known as the Big Bang—a cataclysmic event of unimaginable energy and transformation. From the fiery emergence of the singularity to the gradual formation of galaxies, stars, and planets, our odyssey unveils the epic narrative of cosmic evolution.

Yet, our voyage extends beyond the confines of our home planet. The tantalizing pursuit of habitable worlds, where the conditions for life may exist beyond Earth, becomes our guiding star. We venture into the enigmatic Goldilocks zone, where the delicate balance of environmental factors fosters the potential for life. Our telescopic lenses peer into the vast expanse of exoplanets, each with its own unique narrative waiting to be unraveled.

As intrepid explorers of the cosmic frontier, we confront the intricacies of interstellar communication. We send signals into the cosmic void, harboring the hope of a response that might bridge the unfathomable distances separating us from potential extraterrestrial civilizations. These exchanges of thoughts and ideas, transcending the bounds of our imagination, symbolize the pinnacle of our pursuit.

This expedition is not a solitary endeavor but a collective odyssey that unites us as seekers of truth, understanding, and communion with the cosmos. As we navigate the limitless expanse of space, we recognize that our exploration is an eternal journey—a testament to the boundless human spirit and an enduring tribute to our intrinsic desire to comprehend the universe that envelops us.

The Boundless Frontiers Ahead

As we bring this extraordinary exploration of the cosmos to a contemplative close, we find ourselves at a unique juncture where questions outnumber answers. The mysteries of dark matter and dark energy, the enigmatic prospects of a multiverse, and the ambitious goal of unifying fundamental forces in the universe all stand before us as tantalizing enigmas. Our journey, rather than providing finality, has illuminated the boundless frontiers of science, inviting us to venture further into uncharted territories of knowledge.

In our pursuit of understanding, the cosmic mysteries that have emerged from our exploration serve as beacons guiding our intellectual curiosity. Dark matter, a mysterious substance that neither emits nor absorbs light, challenges our fundamental comprehension of the cosmos. Dark energy, the elusive force responsible for the universe's accelerating expansion, underscores the urgency of unraveling its nature. The contemplation of a multiverse, with its myriad of parallel universes, reshapes our perceptions of reality and existence.

As we stand on the precipice of scientific discovery, we acknowledge that our quest for knowledge is bound neither by time nor space. The horizons of science continue to expand, beckoning us to explore the uncharted territories of quantum gravity, unified theories, and the very fabric of spacetime. These intellectual endeavors remind us that the pursuit of knowledge knows no limits and that the universe is an ever-unfolding tapestry of wonder.

In our relentless pursuit of understanding, we celebrate the inquisitiveness that drives us to explore the unknown. Each question begets another, each mystery deepens our resolve, and each discovery fuels our insatiable hunger for knowledge. As we gaze toward the boundless frontiers ahead, we are reminded that the cosmos is an open book, and our exploration is an eternal odyssey—a testament to the enduring human spirit's capacity to embrace the enigma of the universe.

Humanity's Cosmic Legacy

Space exploration transcends the realm of pure science; it is an embodiment of our intrinsic urge to expand the horizons of the familiar. It serves as a wellspring of inspiration for the generations that follow, igniting their dreams, fostering discovery, and catalyzing innovation. Beyond the scientific frontiers, it introduces us to a realm of ethical contemplation concerning contact with potential extraterrestrial life and the conscientious utilization of cosmic resources, serving as a poignant reminder of our role as stewards of the cosmos.

The act of exploration, whether of our celestial neighbors or distant galaxies, reflects our boundless curiosity—a trait intrinsic to humanity. It is the very same curiosity that spurred us to venture beyond our ancestral homelands, set sail across uncharted seas, and ascend into the skies. Now, we cast our gaze skyward once more, not only as explorers but as custodians of the cosmic heritage bestowed upon us.

Space exploration's resounding legacy extends far beyond the confines of our spacecraft and the boundaries of our solar system. It reverberates through time, echoing in the aspirations of future generations. It serves as a beacon lighting the path for those who dare to dream and empowers them to realize those dreams through the relentless pursuit of knowledge and innovation.

However, as we reach outward to touch the cosmos, we also confront profound ethical considerations. The prospect of contact with extraterrestrial intelligences challenges our perspectives on life, consciousness, and our place in the universe. The responsible use of cosmic resources reminds us of our duty as stewards to safeguard and sustain the delicate equilibrium of our cosmic home.

In closing, the legacy of space exploration is a testament to the enduring spirit of human discovery, an emblem of our unwavering determination, and a symbol of our boundless potential. As we navigate the cosmos and navigate the ethical conundrums it presents, we do so not only for ourselves but for the generations yet to come. In this cosmic odyssey, humanity's legacy is not merely etched in the stars; it is inscribed in the hearts and minds of those who continue to gaze upward and wonder.

Into the Unknown

In the grand tapestry of the universe, our existence registers as but a fleeting moment—a minuscule yet remarkable blip in the cosmic timeline. Yet, within this infinitesimal slice of time, we have demonstrated an insatiable curiosity that knows no bounds, and an unyielding pursuit of knowledge that transcends the limits of our understanding. It is this ceaseless quest that possesses the power to shape the destiny of humanity and define our place in the cosmos.

As we cast our gaze into the limitless expanse of the unknown, we are reminded that the question echoing through the annals of time, "Secrets of the Cosmos" is not a culmination but a commencement—a point of embarkation for an infinite journey of exploration and discovery. This question, rather than signaling an end, heralds the commencement of an odyssey that shall traverse the galaxies, traverse the aeons, and traverse the very essence of existence.

In the cosmic theater, we stand as actors on a stage, momentarily visible against the backdrop of eternity. The mysteries that beckon us are as boundless as the universe itself, spanning the realms of particle physics, quantum mechanics, and the unfathomable recesses of space-time. The enigma of dark matter, the intricacies of interstellar communication, and the tantalizing prospects of extraterrestrial life are but threads in the intricate tapestry of the cosmos, awaiting our unraveling.

As we venture forth into the celestial abyss, we are infused with a sense of purpose—a purpose that transcends the individual, the nation, and the species. It is a purpose that aligns with the very essence of our existence—to explore, to question, to understand. In this boundless expanse, the cosmos extends its invitation, urging us to seek answers, to push boundaries, and to illuminate the mysteries that shroud the universe in profound obscurity.

In closing, the question that resonates within our collective consciousness—whether we are alone in the universe—will continue to echo through the corridors of time. It is a question that shall guide our endeavors, shape our destiny, and inspire future generations. As we journey into the unknown, we do so with an enduring commitment to exploration, an unwavering belief in the boundless potential of human intellect, and a profound understanding that in the cosmic quest for answers, we embark on an infinite voyage—an odyssey that transcends time and transcends space.